Arid and semi-arid regions face major challenges in the management of scarce freshwater resources under pressures of population, economic development, climate change, pollution and over-abstraction. Groundwater is commonly the most important water resource in these areas. Groundwater models are widely used globally to understand groundwater systems and to guide decisions on resource management and protection from pollution. However, the hydrology of arid and semi-arid areas is very different from that of humid regions, and there is little guidance on the special challenges of groundwater modelling for these areas. This book brings together the worldwide experience of internationally leading experts to fill this gap in the scientific and technical literature. It introduces state-of-the-art methods for the modelling of groundwater resources and their protection from pollution. It is illustrated with a wide-ranging set of examples from a variety of regions, including India, China, Africa and the Middle East.

The book is valuable for researchers, practitioners in developed and developing countries, and graduate students in hydrology, hydrogeology, water resources management, environmental engineering and geography.

HOWARD WHEATER is Professor of Hydrology at Imperial College, London. He is past-President of the British Hydrological Society, a Fellow of the Royal Academy of Engineering, a Fellow of the Institution of Civil Engineers, and a life member of the International Water Academy. His research interests are in hydrology and water resources, with wide-ranging applications including climate change, surface and groundwater hydrology, floods, water resources and water quality. He has published over 200 peer-reviewed papers and 6 books. Academic awards include various UK prizes and the 2006 Prince Sultan bin Abdulaziz International Prize for Water. He has been extensively involved in flood and water resource projects in the UK and internationally, providing advice to states and international governments. He has a particular interest in the hydrology of arid areas, and has worked in Oman, Saudi Arabia, Yemen, UAE, Jordan, Syria and Egypt, and Arizona. He chairs UNESCO's G-WADI arid zone water resources programme and was invited by the Japanese government to give a keynote address on water scarcity to the 2003 Kyoto World Water conference.

SIMON MATHIAS holds a Lectureship in the Department of Earth Sciences at Durham University. Prior to this position he was active in the field of groundwater engineering as a researcher and lecturer within the Department of Civil and Environmental Engineering at Imperial College, London. His principal expertise lies in the development of mathematical models to describe flow and transport of reactive contaminants in porous and fractured porous media. Dr Mathias has worked on a broad range of applications including vadose zone transport of nutrients in fracture rock systems, plant uptake of radionuclides, aquifer characterisation studies, buoyancy-driven flow problems, CO_2 geo-sequestration and hydraulic fracture propagation. He has published widely in international peer-reviewed journals. Dr Mathias is an elected committee member of the British Hydrological Society.

XIN LI is Professor at the Cold and Arid Regions Environmental and Engineering Research Institute (CAREERI), Chinese Academy of Sciences (CAS). His primary research interests include land data assimilation, application of remote sensing and GIS in hydrology and cryosphere science, and integrated watershed study. He is currently director of the World Data Center for Glaciology and Geocryology, chair of the working group on remote sensing of the Chinese Committee for WCRP/CliC and IUGG/IACS, co-chair of the working group on theory and method of the China Association for Geographic Information System, and a member of American Geophysical Union. He is also the secretary of G-WADI Asia established by UNESCO IHP. He has published over 120 journal articles and monographs. He was recipient of the Outstanding Science and Technology Achievement Prize of the CAS in 2005, First Class Science and Technology Progress Prize of Gansu Province, and the Seventh National Award for Young Geographers in 2003.

INTERNATIONAL HYDROLOGY SERIES

The **International Hydrological Programme** (IHP) was established by the United Nations Educational, Scientific and Cultural Organization (UNESCO) in 1975 as the successor to the International Hydrological Decade. The long-term goal of the IHP is to advance our understanding of processes occurring in the water cycle and to integrate this knowledge into water resources management. The IHP is the only UN science and educational programme in the field of water resources, and one of its outputs has been a steady stream of technical and information documents aimed at water specialists and decision makers.

The **International Hydrology Series** has been developed by the IHP in collaboration with Cambridge University Press as a major collection of research monographs, synthesis volumes and graduate texts on the subject of water. Authoritative and international in scope, the various books within the series all contribute to the aims of the IHP in improving scientific and technical knowledge of freshwater processes, in providing research knowhow and in stimulating the responsible management of water resources.

Groundwater Modelling in Arid and Semi-Arid Areas

Howard S. Wheater

Imperial College of Science, Technology and Medicine, London

Simon A. Mathias

Durham University

Xin Li

Chinese Academy of Sciences, Lanzhou, China

CAMBRIDGE
UNIVERSITY PRESS

CAMBRIDGE
UNIVERSITY PRESS

The Edinburgh Building, Cambridge CB2 8RU, UK

Published in the United States of America by Cambridge University Press, New York

Cambridge University Press is part of the University of Cambridge.

It furthers the University's mission by disseminating knowledge in the pursuit of
education, learning and research at the highest international levels of excellence.

www.cambridge.org
Information on this title: www.cambridge.org/9781107690110

First published 2010
First paperback edition 2013

A catalogue record for this publication is available from the British Library

Library of Congress Cataloguing in Publication data

Groundwater modelling in arid and semi-arid areas / [edited by] Howard S. Wheater, Simon A. Mathias, Xin Li.
    p. cm. – (International hydrology series)
  ISBN 978-0-521-11129-4 (Hardback)
1. Hydrologic models. 2. Groundwater–Simulation methods. 3. Arid regions. I. Wheater, Howard. II. Mathias, Simon A.
III. Li, Xin. IV. Title. V. Series.
  GB656.2.H9G76 2010
  551.4909154–dc22

                                                              2010024617

ISBN 978-0-521-11129-4 Hardback
ISBN 978-1-107-69011-0 Paperback

Contents

The colour plates will be found between pages 20 and 21.

Contributors

S. Ahmed
Indo-French Centre for Groundwater Research
National Geophysical Research Institute
Hyderabad
India

P. Bauer-Gottwein
Institut for Vand og Miljøteknologi
Danmarks Tekniske Universitet
Bygningstorvet
Bygning 115, rum 154
Kongens Lyngby
Denmark

P. Brunner
Centre of Hydrogeology and
Geothermics (CHYN)
Rue Emile-Argand 11-CP158
CH-2009 Neuchâtel
Switzerland

A. P. Butler
Department of Civil and Environmental Engineering
Imperial College London
London
UK

J. Carrera
Department of Geotechnical Engineering
and Geosciences
Technical University of Catalonia
Barcelona
Spain

W. M. Edmunds
School of Geography and the Environment
University of Oxford
Oxford
UK

T. Graf
Institute of Fluid Mechanics
Department of Civil Engineering
Gottfried Wilhelm Leibniz University
Hannover Appelstrasse 9A
30167 Hannover
Germany

L. Kgotlhang
Tsodilo Resources
Maun
Botswana

W. Kinzelbach
Institute of Environmental Engineering
ETH Zurich
Zurich
Switzerland

H. Kooi
Department of Hydrology and Geo-Environmental Sciences
Faculty of Earth and Life Sciences
VU University Amsterdam
De Boelelaan 1085
Amsterdam
Nederland

L. Li
Centre for Eco-Environmental Modelling
Hohai University
Nanjing
P R China

S. A. Mathias
Department of Earth Sciences
Durham University
Science Laboratories
Durham
UK

C. Milzow
Institut for Vand og Miljøteknologi
Danmarks Tekniske Universitet
Bygningstorvet
Bygning 115, rum 154
Kongens Lyngby
Denmark

A. Nabi
Indo-French Centre for Groundwater Research
National Geophysical Research Institute
Hyderabad
India

S. Owais
Indo-French Centre for Groundwater Research
National Geophysical Research Institute
Hyderabad
India

V. Post
Department of Hydrology and Geo-Environmental Sciences
Faculty of Earth and Life Sciences
VU University Amsterdam
De Boelelaan 1085
Amsterdam
Nederland

H. Prommer
Land and Water – Floreat WA
CSIRO
Underwood Avenue
Floreat Park
Australia

S. Sarah
Indo-French Centre for Groundwater Research
National Geophysical Research Institute
Hyderabad
India

C. T. Simmons
School of the Environment and National Centre for Groundwater
Research and Training
Flinders University
Adelaide
Australia

R. Therrien
Department of Geology and Geological Engineering
Laval University
Quebec City
Canada

A. von Boetticher
Eidg. Forschungsanstalt WSL
Zürcherstrasse 111
Birmensdorf
Switzerland

C. I. Voss
USGS
431 National Center
12201 Sunrise Valley Drive
Reston
Virginia
USA

J. Ward
School of the Environment
Flinders University
Adelaide
Australia

A. Werner
School of Chemistry,
Physics and Earth Sciences
Flinders University
Adelaide
Australia

H. S. Wheater
Department of Civil and Environmental Engineering
Imperial College London
London
UK

Preface

Arid and semi-arid regions present special challenges for water management. They are, by definition, areas where water is at its most scarce, and face great pressures to deliver and manage freshwater resources. Demand for water has increased dramatically, due to population growth, increasing expectations for domestic water use, and expansion of industrial and agricultural activities. Available water resources have been reduced by pollution and over-abstraction. Many of the world's arid regions are further threatened by climate change. In addition, the science base to support water management remains limited. Hydrological processes can be very different from those of humid regions, precipitation and flow exhibit extreme variability in space and time, and data are often restricted in spatial coverage, record length and data quality.

UNESCO has identified, within the International Hydrological Programme, a special need to exchange knowledge on scientific aspects of water resources (with respect to both quantity and quality) in arid and semi-arid lands, and is supporting a number of regional centres to promote exchange of information and dissemination of good practice. At the global level, UNESCO has initiated G-WADI, a Global network for Water and Development Information for arid lands. Information on G-WADI products and a news-watch service can be found on the G-WADI website (www.gwadi.org). G-WADI aims to facilitate the global dissemination of state-of-the-art scientific knowledge and management tools, and to facilitate the sharing of scientific and technical knowledge and management experience of new and traditional technologies to conserve water.

With the support of UNESCO and the UK Government, the first major G-WADI event was held in Roorkee, India, in March 2005, focusing on the surface-water modelling tools required to support water management in arid and semi-arid areas. The strategy was to bring together the world's leading experts to provide lectures and tutorials on this topic. This resulted in a book in the International Hydrology Series, *Hydrological Modelling in Arid and Semi-Arid Areas*, published by Cambridge University Press in late 2007.

Groundwater is commonly the most important water resource in arid areas, but is particularly difficult to quantify in terms of sustainability, and almost universally suffers from problems of over-abstraction, declining water tables and degradation of water quality. Hence, a second event was held in Lanzhou, China, in 2007, focusing on groundwater modelling in arid and semi-arid areas, and attended by 56 participants from 22 countries. The aims of the workshop were:

1. to bring together the world's leading experts in arid zone groundwater modelling to deliver a definitive set of lectures and case studies to an audience of active researchers from the world's arid regions;
2. to draw on the experience of the workshop participants in developing this material and to consider recommendations for future activities;
3. to make this material available to the global community through UNESCO and in particular the G-WADI website (www.gwadi.org);
4. to stimulate follow-up activities, regionally and globally.

The material from this workshop provides the content of this book. It brings together state-of-the-art information on groundwater data, modelling and management, specifically focused on the challenges of arid and semi-arid areas, and we can say with confidence that the authors represent some of the world's most distinguished authorities.

The structure of the book is as follows:

Chapter 1 describes the context for groundwater management in arid and semi-arid areas, including historical development and current pressures, and the associated needs for modelling, and provides a summary of the book content and structure. Chapter 2 provides a review of some of the special hydrological features of arid areas, with a particular focus on groundwater recharge processes, and examples from the Arabian peninsula. Chapter 3 introduces isotopic and geochemical methods as important sources of information and insight into groundwater systems, with case studies from Africa and the Mediterranean. Chapter 4 provides an overview of groundwater modelling, including the treatment of spatial variability, calibration and uncertainty, and Chapter 5 illustrates the application of geostatistics to an Indian case study. Groundwater source protection is discussed in Chapter 6, and Chapter 7 provides a comprehensive discussion of the problems of density-dependent groundwater flows associated with salinity effects. Finally, Chapter 8 addresses sustainable water management in arid and semi-arid regions, with case studies from North Africa, Southern Africa and China.

Acknowledgements

The editors particularly wish to thank the contributors to the book for their enthusiastic input to both the workshop and the book, the sponsors of the workshop, without whom none of this activity would have been possible, and the workshop attendees, who provided an informed audience and helpful feedback. Financial support for the contributors, regional representatives and international organisation was provided by UNESCO's International Hydrology Programme and the UK Government's Department for International Development. We are indebted to the Cold and Arid Regions Environmental and Engineering Research Institute, Lanzhou, who provided local organisation and superb hospitality. Support for participants from the Asian region to attend the workshop was provided by UNESCO's regional offices in Delhi and Tehran.

1 Groundwater modelling in arid and semi-arid areas: an introduction

S. A. Mathias and H. S. Wheater

1.1 INTRODUCTION

Water resources globally face unprecedented challenges, but these are at their greatest in the world's arid and semi-arid regions. Recent IPCC estimates (Kundzewicz et al., 2007) state that between 1.4 and 2.1 billion people live in areas of water stress; those numbers are expected to increase significantly under the pressures of population growth and climate change.

By definition, arid and semi-arid regions have limited natural water resources, and precipitation and runoff have very high variability in space and time. Traditional societies recognised these characteristics and developed sustainable water management solutions. In higher rainfall areas, for example in the mountains of northern Yemen and Greek islands such as Cephalonia, rainwater was harvested from roofs and paved surfaces and stored for household or community use. In desert areas, such as Arabia's 'empty quarter', with infrequent, spatially localised rainfall, nomadic communities would follow rainfall occurrence, using water from surface storage or shallow groundwater for a few months to support themselves and their livestock, before moving on. For agriculture, terraced systems were developed to focus infiltration to provide soil moisture for crop water needs (as in the mountains of Northern Oman and Yemen), and earth dams were built to divert flash floods onto agricultural land for surface irrigation (as in South West Saudi Arabia). In parts of the Middle East, groundwater was extracted sustainably using qanats (Iran) or afalaj (Oman), ancient systems of tunnels for gravity drainage of groundwater, developed over centuries or longer.

The twentieth century has seen pressures of increased population, increased social expectations for domestic use and increased agricultural water use, and the general result has been unsustainable use of water, from the developed and wealthy economies of Southwestern USA, to poorer countries of South America, Africa, Asia and the Middle East. This has led to declining groundwater levels, reduced (or non-existent) surface-water flows and loss of wetlands, such as the Azraq oasis in North East Jordan – a RAMSAR wetland site now totally dry due to over-pumping of groundwater. There is a range of adverse effects associated with this over-exploitation, for example, loss of important habitats, deteriorating water quality, including ingress of saline water in coastal aquifers, and land subsidence. In the twenty-first century we face the added challenge of climate change. It is essential therefore that we recognise the need for sustainable use of water, balancing the long-term use with the long-term water availability, and learn to use best practice from traditional and new methods of water management. This requires social recognition of water scarcity, the political will to confront difficult societal choices, and good science and engineering to develop and support sustainable management solutions.

This book has been developed under UNESCO's G-WADI programme and aims to provide state-of-the-art guidance for those involved in water science and management in arid and semi-arid areas concerning the modelling methods that are needed to characterise water resource systems and their sustainable yield, and to provide protection from pollution. In an earlier book (Wheater et al., 2008), G-WADI mainly addressed the issues of surface-water systems. Here we consider groundwater.

1.2 GROUNDWATER RESOURCES AND MODELLING NEEDS IN ARID AND SEMI-ARID AREAS

Groundwater is commonly the most important water resource in arid areas. In areas of high evaporation and limited rainfall, groundwater provides natural storage of water which is protected from surface evaporation, it is spatially distributed, and it can be developed with limited capital expenditure. Groundwater also provides a potential storage that can be managed to increase the useful water resource. One example is the use of recharge

Groundwater Modelling in Arid and Semi-Arid Areas, ed. Howard S. Wheater, Simon A. Mathias and Xin Li. Published by Cambridge University Press.

dams, as developed on the coastal (Batinah) plain of the Sultanate of Oman, to temporarily retain flash floods and focus the infiltration of surface water to recharge the underlying aquifer system. A recent review of aspects of Managed Aquifer Recharge (MAR) can be found in Dillon *et al.*, (2009).

However, as noted above, in all arid regions, groundwater resources are under threat from over-abstraction and pollution. The widescale deployment of powerful motorised pumps, in the absence of effective regulation, has led to major problems of resource depletion, declining water levels and deterioration of water quality. A particular and widespread problem arises where over-abstraction in coastal aquifers leads to ingress of salt water, which can make the resource unusable (see Chapter 7). And the time-scales of groundwater response are such that pollution may affect a resource for decades or centuries, if indeed full recovery is possible. Moreover, in some areas much of the water being abstracted from deep aquifers is non-renewable, being a legacy from wetter climates in the past.

Management of aquifers must make best use of available information and balance competing pressures. However, the ability to observe the subsurface is limited, and hence characterisation of available groundwater resources, and in particular the natural recharge that sustains these resources, is difficult and often highly uncertain. Models are therefore used worldwide as a tool to assimilate data on groundwater systems and to guide decisions on management strategies for groundwater resources and protection from, and remediation of, pollution. It is the aim of this book to provide insight into the modelling process, the integration of data sources and the issues of sustainable management of groundwater in the context of the special needs of arid and semi-arid areas.

1.3 THE BOOK CONTENT AND STRUCTURE

Chapter 2 summarises current knowledge of hydrological processes in arid and semi-arid areas, including rainfall, runoff, surface-water–groundwater interactions (including wadi-bed transmission losses) and groundwater recharge, drawing on examples from the Middle East, southern Africa and the USA. There is limited guidance in the scientific and technical literature on the hydrology of arid areas or on the modelling tools needed to underpin groundwater management.

The hydrology of arid areas is very different from that of humid areas, and Chapter 2 outlines the special technical challenges in the assessment and management of water. Precipitation is a particular problem. It is generally characterised by very high variability in space and time; runoff from a single large storm can exceed the total runoff from a sequence of years, and intense rainfall can be spatially localised. Flows are infrequent

but times to peak are typically short and discharges can be large. Flows often decrease with distance downstream as water is lost to channel bed infiltration. Hydrological data from such areas are difficult to capture and available records are limited in length and quality. Rainfall and flood events are thus difficult to quantify, and this leads to difficulty in rainfall–runoff modelling and in the estimation of groundwater recharge. It is argued that distributed modelling is needed to capture the spatial variability of precipitation, runoff and recharge processes, to quantify groundwater recharge and to assess the potential for recharge management. Given the lack of data, integrated modelling of surface and groundwater systems is needed to assimilate the available data and inform management decisions.

Chapter 3 introduces the use of isotopic and geochemical methods in the analysis of groundwater systems in arid and semi-arid areas. This includes methods of investigation, time-scales and palaeohydrology, rainfall and unsaturated zone chemistry, and the hydrochemistry of groundwater systems in arid and semi-arid areas. Chemical and especially isotopic methods can provide valuable insights into groundwater systems and their functioning. The development of water quality may be viewed as an evolution over time, for example: (i) undisturbed evolution under natural conditions; (ii) the development phase with some disturbance of natural conditions, especially stratification; (iii) development with contamination; and (iv) artificially managed systems. Hence the methods can identify the interface between modern water and palaeowaters as a basis for sustainable management of water resources. Chapter 3 focuses on the use of isotopic and geochemical methods to assist the development of conceptual models for recharge sequences. The application of hydrogeochemical techniques in understanding the water quality problems of semi-arid regions is described and follows the chemical pathway of water from rainfall through the hydrological cycle. A number of case studies from Sudan, Nigeria and Senegal as well as North African counties and China are used to illustrate the main principles.

Chapter 4 sets out the basic strategy for modelling groundwater flow and transport processes, including modelling objectives and modelling procedure (including conceptualisation, calibration and error analysis, model selection, predictions and uncertainty), and introduces advanced methods to represent spatial variability (kriging, conditional simulation, sequential Gaussian simulation and facies simulation), and advanced methods of calibration and uncertainty propagation. Groundwater modelling involves a range of complex issues from understanding site geology to the development and application of suitable numerical methods. Chapter 4 has been arranged to cover the basics of these while directing the reader to specialised literature for details. The chapter starts by describing the flow phenomenon in terms of equations and basic numerical methods. This is followed by a discussion on the general procedure of building models using numerical codes. Since heterogeneity is such a ubiquitous problem, the chapter dedicates a large section to describing many of the basic geostatistical and upscaling tools available to deal with spatial variability. From here, a formalisation of the calibration procedure is made. Finally, the chapter concludes with a discussion on methods for uncertainty assessment.

In most situations, observed groundwater and associated data are sparsely distributed within the area of interest. The study of regionalised variables starts from the ability to interpolate a given field from only a limited number of observations while preserving the theoretical spatial correlation. This is often accomplished by means of a geostatistical technique called kriging. Compared with other interpolation techniques, kriging is advantageous because it considers the number and spatial configuration of the observation points, the position of the data point within the region of interest, the distances between the data points with respect to the area of interest and the spatial continuity of the interpolated variable. Chapter 5 further describes the theory of regionalised variables and how it can be applied to groundwater modelling studies. The discussed techniques are then demonstrated through a case study from the Maheshwaram watershed of Andhra Pradesh, India.

As discussed above, groundwater resources are vulnerable to pollution. This may be either from point sources, for example illegal or accidental discharges from industry or domestic wastes, or from diffuse pollution, as in the case of nutrient pollution from agriculture. It is a characteristic of groundwater systems that remediation of pollution is difficult, costly and time-consuming, and may only be partially successful. There are therefore strong societal imperatives to protect groundwater resources. Chapter 6 addresses this issue. First, the techniques available for the assessment of aquifer vulnerability are presented, considering their relative advantages and disadvantages, and their applications to the particular characteristics of arid and semi-arid areas. Second, the definition of protection zones for individual water sources is discussed. This requires definition of either a total catchment area for a well, or a time-related capture zone, i.e. defining the time of travel of potential pollutants to a given source. Again, applications to arid and semi-arid areas are discussed, including detailed methodologies to specify flow pathways and pollutant travel times, and simpler rules of thumb relevant, for example, to the relative siting of wells and pit latrines.

Arid and semi-arid climates are mainly characterised as those areas where precipitation is less (and often considerably less) than potential evapotranspiration. These climate regions are ideal environments for salt to accumulate in natural soil and groundwater settings since evaporation and transpiration remove freshwater from the system, leaving residual salts behind. Similarly, the characteristically low precipitation rates reduce the potential for salt to be diluted by rainfall. Salt flats, playas, sabkhas and

saline lakes, for example, are therefore ubiquitous features of arid and semi-arid regions throughout the world. In such settings, variable density flow phenomena are expected to be important, especially where hypersaline brines overlie less dense groundwater at depth. In coastal regions, sustainable management of groundwater resources must also take into account the variable density phenomena of seawater intrusion. Chapter 7 provides a comprehensive state-of-the-art review of the issues of modelling density-dependent flows. The importance of density-dependent flows for arid regions is discussed, variable density physics is introduced, the relevant governing equations for simulating variable density flow are presented and various commercial numerical codes are compared. Model benchmarking is discussed in detail. Applications and case studies include seawater intrusion and tidally induced phenomena, transgression–regression salinisation of coastal aquifers, aquifer storage and recovery, fractured rock flow and chemically reactive transport modelling.

As a result of the ever-growing global population, pressure on water resources is increasing continually. In many cases, the presently applied management practices are non-sustainable and lead to serious water-related problems such as the depletion of aquifers and the accumulation of substances to harmful levels, as well as to water conflicts or economically infeasible costs. Chapter 8, as the final chapter, addresses sustainable water management in arid and semi-arid regions. Associated problems are poignantly illustrated by case studies. The northwest Sahara aquifer system is used to show the consequences of the overpumping of aquifers, and a typical upstream–downstream problem is discussed with the example of the Okavango delta. Another case study discussed is the Yanqi basin in China. The Yanqi basin is a typical example showing how inappropriate irrigation practices can lead to soil salinisation and ecological problems downstream. This final chapter clearly demonstrates the important role numerical modelling can play as a tool to develop sustainable groundwater management practices. Some of the most common problems in setting up reliable models are highlighted and ideas on how to address these problems are given. Useful guidance is also given to help narrow the often prevalent gap between scientists and decision makers.

1.4 CONCLUSIONS

Groundwater lies at the heart of many of the water management issues faced in arid and semi-arid regions. In this book we attempt to provide some insight into the tools and techniques available to support the sustainable management of groundwater resources. Assessment of recharge is fundamental to the definition of sustainable groundwater yields, yet this remains a very challenging area, particularly for arid and semi-arid areas. We present isotopic and geochemical methods that can be used for recharge estimation, and argue that modelling provides an essential set of tools for the integration of information on groundwater systems. We provide a state-of-the-art assessment of groundwater flow and quality modelling methods, with recent developments in the representation of uncertainty, including geostatistical methods. We provide methodologies for the protection of groundwater resources, through aquifer vulnerability assessment and the definition of well-capture areas. In dry areas, salinity is commonly a major issue, not only for saline intrusion in coastal aquifers, but more generally for water bodies in a high-evaporation environment. The implications for modelling and analysis can be profound, and we present a comprehensive discussion of the associated modelling capabilities and some outstanding research challenges. The book draws on a wide range of case study examples to illustrate the methods presented, and concludes with a final chapter that discusses the practical issues of modelling for decision support, and the important role of models as a means of communication between different stakeholders in the water management arena.

REFERENCES

Dillon, P., Kumar, A., Kookana R. *et al.* (2009) *Managed Aquifer Recharge – Risks to Groundwater Dependent Ecosystems – A Review*. Water for a Healthy Country Flagship Report to Land & Water Australia. May 2009, CSIRO.

Kundzewicz, Z. W., Mata, L. J., Arnell, N. W. *et al.* (2007) Freshwater resources and their management. In *Climate Change 2007: Impacts, Adaptation and Vulnerability. Contribution of Working Group II to the Fourth Assessment Report of the Intergovernmental Panel on Climate Change*, ed. M. L. Parry, O. F. Canziani, J. P. Palutikof, P. J. van der Linden and C. E. Hanson, 173–210. Cambridge University Press.

Wheater, H. S., Sorooshian, S. and Sharma, K. D. (eds.) (2008) *Hydrological Modelling in Arid and Semi-Arid Areas*. Cambridge University Press.

2 Hydrological processes, groundwater recharge and surface-water/groundwater interactions in arid and semi-arid areas

H. S. Wheater

2.1 GROUNDWATER RESOURCES, GROUNDWATER MODELLING AND THE QUANTIFICATION OF RECHARGE

The traditional development of water resources in arid areas has relied heavily on the use of groundwater. Groundwater uses natural storage, is spatially distributed and, in climates where potential evaporation rates can be of the order of metres per year, provides protection from the high evaporation losses experienced by surface-water systems. Traditional methods for the exploitation of groundwater have been varied, including the use of very shallow groundwater in seasonally replenished riverbed aquifers (as in the sand rivers of Botswana), the channelling of unconfined alluvial groundwater in afalaj (or qanats) in Oman and Iran, and the use of hand-dug wells. Historically, abstraction rates were limited by the available technology, and rates of development were low, so that exploitation was generally sustainable.

However, in recent decades, pump capacities have dramatically increased and hence agricultural use of water has grown rapidly, while the increasing concentration of populations in urban areas has meant that large-scale well fields have been developed for urban water supply. A common picture in arid areas is that groundwater levels are in rapid decline; in many instances this is accompanied by decreasing water quality, particularly in coastal aquifers where saline intrusion is a threat. Associated with population growth, economic development and increased agricultural intensification, pollution has also become an increasing problem. The integrated assessment and management of groundwater resources is essential so that aquifer systems can be protected from pollution and over-exploitation. This requires the use of groundwater models as a decision support tool for groundwater management.

Some of the most difficult aspects of groundwater modelling concern the interaction between surface-water and groundwater systems. This is most obviously the case for the quantification of long-term recharge, which ultimately defines sustainable yields. Quantification of recharge remains the major challenge for groundwater development worldwide, but is a particular difficulty in arid areas where recharge rates are small, both as a proportion of the water balance and in absolute terms. However, more generally, the interactions between surface-water and groundwater systems are important. In arid areas, infiltration from surface-water channels, as a 'transmission loss' for surface flows, may be a major component of groundwater recharge, and there is increasing interest in the active management of this process to focus recharge (for example, in an extensive programme of construction of 'recharge dams' in northern Oman). Conversely, the discharge of groundwater to surface-water systems can be important in terms of valuable ecosystems. Hence the main thrust of this chapter is to review hydrological processes in arid and semi-arid areas, to provide the context for surface/groundwater interactions, and their analysis and modelling.

2.2 HYDROLOGICAL PROCESSES IN ARID AREAS

Despite the critical importance of water in arid and semi-arid areas, hydrological data have historically been severely limited. It has been widely stated that the major limitation of the development of arid zone hydrology is the lack of high-quality observations (McMahon, 1979; Nemec and Rodier, 1979; Pilgrim et al., 1988). There are many good reasons for this. Populations are usually sparse and economic resources limited; in addition the climate is harsh and hydrological events infrequent but damaging. However, in the general absence of reliable long-term data and experimental research, there has been a tendency to rely on humid zone experience and modelling tools, and data from other regions. At best, such results will be highly inaccurate. At worst,

Groundwater Modelling in Arid and Semi-Arid Areas, ed. Howard S. Wheater, Simon A. Mathias and Xin Li. Published by Cambridge University Press.

Table 2.1. *Summary of Muscat rainfall data (1893–1959) (after Wheater and Bell, 1983).*

Monthly rainfall (mm)	Jan.	Feb.	Mar.	Apr.	May	June	July	Aug.	Sept.	Oct.	Nov.	Dec.
Mean	31.2	19.1	13.1	8.0	0.38	1.31	0.96	0.45	0.0	2.32	7.15	22.0
Standard deviation	38.9	25.1	18.9	20.3	1.42	8.28	4.93	2.09	0.0	7.62	15.1	35.1
Max.	143.0	98.6	70.4	98.3	8.89	64.0	37.1	14.7	0.0	44.5	77.2	171.2
Mean number of raindays	2.03	1.39	1.15	0.73	0.05	0.08	0.10	0.07	0.0	0.13	0.51	1.6
Max. daily fall (mm)	78.7	57.0	57.2	51.3	8.9	61.5	30.0	10.4	0.0	36.8	53.3	57.2
Number of years on record	63	64	62	63	61	61	60	61	61	60	61	60

there is a real danger of adopting inappropriate management solutions which ignore the specific features of dryland response.

Despite the general data limitations, there has been some substantial and significant progress in development of national data networks and experimental research. This has given new insights, and we can now see with greater clarity the unique features of arid zone hydrological systems and the nature of the dominant hydrological processes. This provides an important opportunity to develop methodologies for flood and water resource management which are appropriate to the specific hydrological characteristics of arid areas and the associated management needs, and hence to define priorities for research and hydrological data. The aim here is to review this progress and the resulting insights, and to consider some of the implications.

2.2.1 Rainfall

Rainfall is the primary hydrological input, but rainfall in arid and semi-arid areas is commonly characterised by extremely high spatial and temporal variability. The temporal variability of point rainfall is well known. Although most records are of relatively short length, a few are available from the nineteenth century. For example, Table 2.1 presents illustrative data from Muscat (Sultanate of Oman) (Wheater and Bell, 1983), which shows that a wet month is one with one or two raindays. Annual variability is marked and observed daily maxima can exceed annual rainfall totals.

For spatial characteristics, information is much more limited. Until recently, the major source of detailed data has been from Southwestern USA, most notably the two relatively small, densely instrumented basins of Walnut Gulch, Arizona (150 km^2), and Alamogordo Creek, New Mexico (174 km^2), established in the 1950s (Osborn *et al.*, 1979). The dominant rainfall for these basins is convective; at Walnut Gulch 70% of annual rainfall occurs from purely convective cells, or from convective cells developing along weak, fast-moving cold fronts, and falls in the period July to September (Osborn and Reynolds, 1963). Raingauge densities were increased at Walnut Gulch to give improved definition of detailed storm structure and are currently better than 1 per 2 km^2. This has shown highly localised rainfall

occurrence, with spatial correlations of storm rainfall of the order of 0.8 at 2 km separation, but close to zero at 15–20 km spacing. Osborn *et al.* (1972) estimated that to observe a correlation of $r^2 = 0.9$ raingauge spacings of 300–500 m would be required.

Recent work has considered some of the implications of the Walnut Gulch data for hydrological modelling. Michaud and Sorooshian (1994) evaluated problems of spatial averaging for rainfall-runoff modelling in the context of flood prediction. Spatial averaging on a 4 km × 4 km pixel basis (consistent with typical weather radar resolution) gave an underestimation of intensity and led to a reduction in simulated runoff of on average 50% of observed peak flows. A sparse network of raingauges (1 per 20 km^2), representing a typical density of flash flood warning systems, gave errors in simulated peak runoff of 58%. Evidently there are major implications for hydrological practice; we will return to this issue below.

The extent to which this extreme spatial variability is characteristic of other arid areas has been uncertain. Anecdotal evidence from the Middle East underlays comments that spatial and temporal variability was extreme (Food and Agriculture Organization, 1981), but data from southwest Saudi Arabia obtained as part of a five-year intensive study of five basins (Saudi Arabian Dames and Moore, 1988) undertaken on behalf of the Ministry of Agriculture and Water, Riyadh, have provided a quantitative basis for assessment. The five study basins range in area from 456 to 4930 km^2 and are located along the Asir escarpment (Figure 2.1), three draining to the Red Sea and two to the interior towards the Rub al Khali. The mountains have elevations of up to 3000 m above sea level (a.s.l.); hence the basins encompass a wide range of altitude, which is matched by a marked gradient in annual rainfall from 30–100 mm on the Red Sea coastal plain to up to 450 mm at elevations in excess of 2000 m a.s.l.

The spatial rainfall distributions are described by Wheater *et al.* (1991a). The extreme spottiness of the rainfall is illustrated for the 2869 km^2 Wadi Yiba by the frequency distributions of the number of gauges at which rainfall was observed given the occurrence of a catchment rainday (Table 2.2). Typical intergauge spacings were 8–10 km, and on 51% of raindays only one or two raingauges out of 20 experienced rainfall. For the more widespread events, subdaily rainfall showed an even more spotty

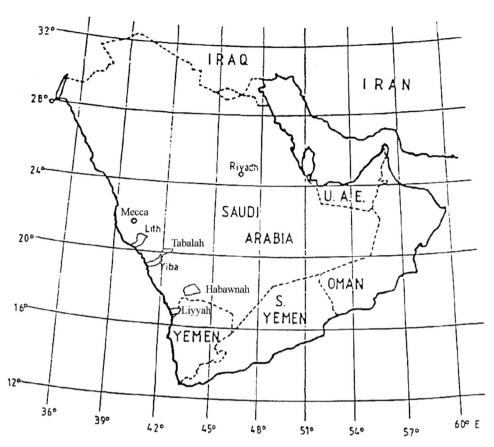

Figure 2.1. Location of Saudi Arabian study basins.

Table 2.2. *Wadi Yiba raingauge frequencies and associated conditional probabilities for catchment rainday occurrence.*

Number of gauges	Occurrence	Probability
1	88	0.372
2	33	0.141
3	25	0.106
4	18	0.076
5	10	0.042
6	11	0.046
7	13	0.055
8	6	0.026
9	7	0.030
10	5	0.021
11	7	0.030
12	5	0.021
13	3	0.013
14	1	0.004
15	1	0.004
16	1	0.005
17	1	0.004
18	1	0.004
19	0	0.0
20	0	0.0
TOTAL	235	1.000

picture than the daily distribution. An analysis of relative probabilities of rainfall occurrence, defined as the probability of rainfall occurrence for a given hour at station B given rainfall at station A, gave a mean value of 0.12 for Wadi Yiba, with only 5% of values greater that 0.3. The frequency distribution of rainstorm durations shows a typical occurrence of one- or two-hour duration point rainfalls, and these tend to occur in mid to late afternoon. Thus rainfall will occur at a few gauges and die out, to be succeeded by rainfall in other locations. This is illustrated for Wadi Lith in Figure 2.2, which shows the daily rainfall totals for the storm of 16 May 1984 (Figure 2.2a), and the individual hourly depths (Figures 2.2b–e). In general, the storm patterns appear to be consistent with the results from Southwest USA, and area reduction factors were also generally consistent with results from that region (Wheater *et al.*, 1989).

The effects of elevation were investigated, but no clear relationship could be identified for intensity or duration. However, a strong relationship was noted between the frequency of raindays and elevation. It was thus inferred that, once rainfall occurred, its point properties were similar over the catchment, but occurrence was more likely at the higher elevations. It is interesting to note that a similar result has emerged from an analysis of rainfall in Yemen (UNDP, 1992), in which it was concluded that daily rainfalls observed at any location are effectively samples from a population that is independent of position or altitude.

It is dangerous to generalise from samples of limited record length, but it is clear that most events observed by those networks are characterised by extremely spotty rainfall, so much so that in the Saudi Arabian basins there were examples of wadi flows generated from zero observed rainfall. However, there were also some indications of a small population of more widespread rainfalls, which would obviously be of considerable importance in terms of surface flows and recharge. This reinforces the need for long-term monitoring of experimental networks to characterise spatial variability.

For some other arid or semi-arid areas, rainfall patterns may be very different. For example, data from arid New South Wales, Australia, have indicated spatially extensive low-intensity rainfalls (Cordery *et al.*, 1983), and recent research in the Sahelian zone of Africa has also indicated a predominance of widespread rainfall. This was motivated by concern to develop improved understanding of land-surface processes for climate studies and modelling, which led to a detailed (but relatively short-term) international experimental programme, the HAPEX-Sahel project

based on Niamey, Niger (Goutorbe *et al.*, 1997). Although designed to study land surface/atmosphere interactions, rather than as an integrated hydrological study, it has given important information. For example, Lebel *et al.* (1997) and Lebel and Le Barbe (1997) note that a 100 raingauge network was installed and report information on the classification of storm types, spatial and temporal variability of seasonal and event rainfall, and storm movement. It was found that 80% of total seasonal rainfall fell as widespread events which covered at least 70% of the network. The number of gauges allowed the authors to analyse the uncertainty of estimated areal rainfall as a function of gauge spacing and rainfall depth.

Recent work in southern Africa (Andersen *et al.*, 1998; Mocke, 1998) has been concerned with rainfall inputs to hydrological models to investigate the resource potential of the sand rivers of northeast Botswana. Here, annual rainfall is of the order of 600 mm, and available rainfall data is spatially sparse and apparently highly variable but of poor data quality. Investigation of the representation of spatial rainfall for distributed water resource modelling showed that use of conventional methods of spatial weighting of raingauge data, such as Theissen

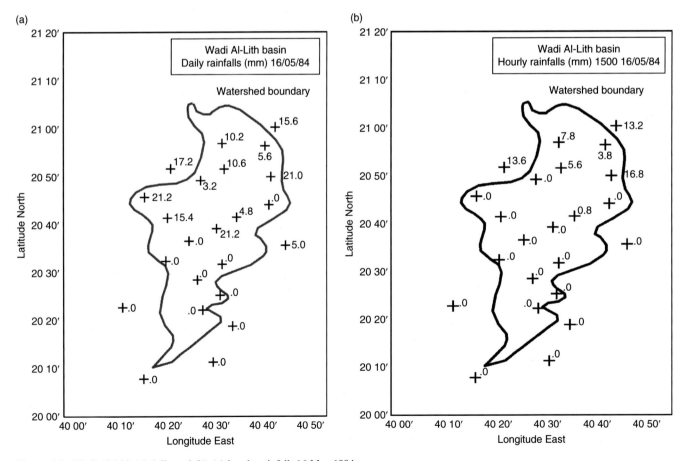

Figure 2.2. Wadi Al-Lith (a) daily and (b)–(e) hourly rainfall, 16 May 1984.

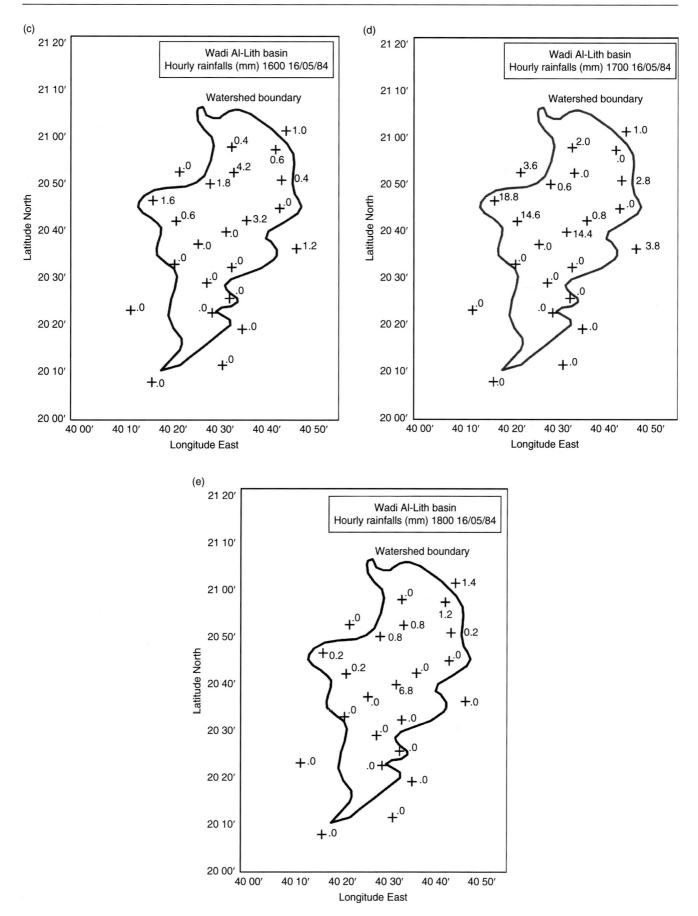

Figure 2.2. (cont.)

polygons, could give large errors. Large sub-areas had rainfall defined by a single possibly inaccurate gauge. A more robust representation resulted from assuming catchment-average rainfall to fall uniformly, but the resulting accuracy of simulation was still poor. Al-Qurashi *et al.* (2008) report application of the KINEROS2 rainfall-runoff model to a catchment in Oman. Rainfall spatial variability was a dominant influence on performance, and McIntyre *et al.* (2007) show that simple empirical relationships can be at least as successful in representing hydrograph characteristics.

2.2.2 Rainfall-runoff processes

The lack of vegetation cover in arid and semi-arid areas removes protection of the soil from raindrop impact, and soil crusting has been shown to lead to a large reduction in infiltration capacity for bare soil conditions (Morin and Benyamini, 1977); hence infiltration of catchment soils can be limited. In combination with the high-intensity short-duration convective rainfall discussed above, extensive overland flow can be generated. This overland flow, concentrated by the topography, converges on the wadi channel network, with the result that a flood flow is generated. However, the runoff generation process due to convective rainfall is likely to be highly localised in space, reflecting the spottiness of the spatial rainfall fields, and to occur on only part of a catchment, as illustrated above.

Linkage between inter-annual variability of rainfall, vegetation growth and runoff production may occur. Our modelling in Botswana suggests that runoff production is lower in a year which follows a wet year, due to enhanced vegetation cover, which supports observations reported by Hughes (1995).

Commonly, flood flows move down the channel network as a flood wave, moving over a bed that either is initially dry or has a small initial flow. Hydrographs are typically characterised by extremely rapid rise times of as little as 15–30 minutes (Figure 2.3). However, losses from the flood hydrograph through bed infiltration are an important factor in reducing the flood volume as the flood moves downstream. These transmission losses dissipate the flood and obscure the interpretation of observed hydrographs. It is not uncommon for no flood to be observed at a gauging station, when further upstream a flood has been generated and lost to bed infiltration.

As noted above, the spotty spatial rainfall patterns observed in Arizona and Saudi Arabia are extremely difficult, if not impossible, to quantify using conventional densities of raingauge network. This, taken in conjunction with the flood transmission losses, means that conventional analysis of rainfall-runoff relationships is problematic, to say the least. Wheater and Brown (1989) present an analysis of Wadi Ghat, a 597 km^2 subcatchment of Wadi Yiba, one of the Saudi Arabian basins discussed

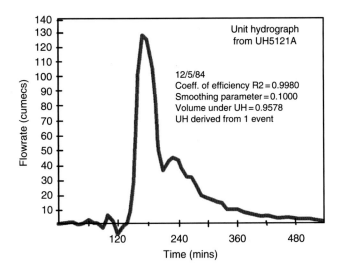

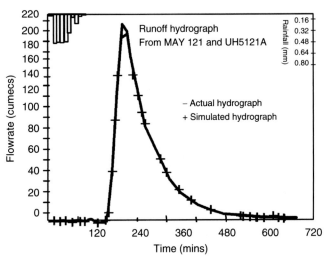

Figure 2.3. Surface-water hydrographs, Wadi Ghat 12 May 1984: observed hydrograph and unit hydrograph simulation.

above. Areal rainfall was estimated from five raingauges and a classical unit hydrograph analysis was undertaken. A striking illustration of the ambiguity in observed relationships is the relationship between observed rainfall depth and runoff volume (Figure 2.4). Runoff coefficients ranged from 5.9 to 79.8%, and the greatest runoff volume was apparently generated by the smallest observed rainfall! Goodrich *et al.* (1997) show that the combined effects of limited storm areal coverage and transmission loss give important differences from more humid regions. Whereas generally basins in more humid climates show increasing linearity with increasing scale, the response of Walnut Gulch becomes more non-linear with increasing scale. It is argued that this will give significant errors in application of rainfall depth–area–frequency relationships beyond the typical area of storm coverage, and that channel routing and transmission loss must be explicitly represented in watershed modelling.

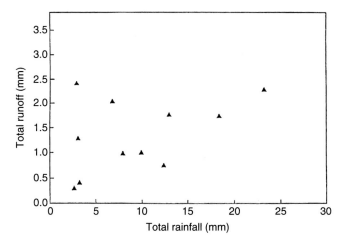

Figure 2.4. Storm runoff as a function of rainfall, Wadi Ghat.

The transmission losses from the surface-water system are a major source of potential groundwater recharge. The characteristics of the resulting groundwater resource will depend on the underlying geology, but bed infiltration may generate shallow water tables, within a few metres of the surface, which can sustain supplies to nomadic people for a few months (as in the Hesse of the north of southern Yemen), or recharge substantial alluvial aquifers with potential for continuous supply of major towns (as in northern Oman and southwest Saudi Arabia).

The balance between localised recharge from bed infiltration and diffuse recharge from rainfall infiltration of catchment soils will vary greatly depending on local circumstances. However, soil moisture data from Saudi Arabia (Macmillan, 1987) and Arizona (Liu *et al.*, 1995), for example, show that most of the rainfall falling on soils in arid areas is subsequently lost by evaporation. Methods such as the chloride profile method (e.g. Bromley *et al.*, 1997) and isotopic analyses (Allison and Hughes, 1978) have been used to quantify the residual percolation to groundwater in arid and semi-arid areas (see Chapter 3 for a fuller discussion of isotopic and geochemical methods).

In some circumstances runoff occurs within an internal drainage basin, and fine deposits can support widespread surface ponding. A well-known large-scale example is the Azraq oasis in northeast Jordan, but small-scale features (Qaa's) are widespread in that area. Small-scale examples were found in the HAPEX Sahel study (Desconnets *et al.*, 1997). Infiltration from these areas is in general not well understood but may be extremely important for aquifer recharge. Desconnets *et al.* (1997) report aquifer recharge of between 5% and 20% of basin precipitation for valley bottom pools, depending on the distribution of annual rainfall.

The characteristics of the channelbed infiltration process are discussed in the following section. However, it is clear that the surface hydrology generating this recharge is complex and extremely difficult to quantify using conventional methods of analysis.

2.2.3 Wadi bed transmission losses

Wadi bed infiltration has an important effect on flood propagation, but also provides recharge to alluvial aquifers. The balance between distributed infiltration from rainfall and wadi bed infiltration is obviously dependent on local conditions, but soil moisture observations from southwest Saudi Arabia imply that, at least for frequent events, distributed infiltration of catchment soils is limited, and that increased near-surface soil moisture levels are subsequently depleted by evaporation. Hence wadi bed infiltration may be the dominant process of groundwater recharge. As noted above, depending on the local hydrogeology, alluvial groundwater may be a readily accessible water resource. Quantification of transmission loss is thus important but raises a number of difficulties.

One method of determining the hydraulic properties of the wadi alluvium is to undertake infiltration tests. Infiltrometer experiments give an indication of the saturated hydraulic conductivity of the surface. However, if an infiltration experiment is combined with measurement of the vertical distribution of moisture content, for example using a neutron probe, inverse solution of a numerical model of unsaturated flow can be used to identify the unsaturated hydraulic conductivity relationships and moisture characteristic curves. This is illustrated for the Saudi Arabian Five Basins Study by Parissopoulos and Wheater (1992a).

In practice, spatial heterogeneity will introduce major difficulties to the upscaling of point profile measurements. The presence of silt lenses within the alluvium was shown to have important effects on surface infiltration as well as subsurface redistribution (Parissopoulos and Wheater, 1990), and subsurface heterogeneity is difficult and expensive to characterise. In a series of two-dimensional numerical experiments it was shown that 'infiltration opportunity time', i.e. the duration and spatial extent of surface wetting, was more important than high flow stage in influencing infiltration, that significant reductions in infiltration occur once hydraulic connection is made with a water table, and that hysteresis effects were generally small (Parissopoulos and Wheater, 1992b). Also sands and gravels appeared effective in restricting evaporation losses from groundwater (Parissopoulos and Wheater, 1991).

Additional process complexity arises, however. General experience from the Five Basins Study was that wadi alluvium was highly transmissive, yet observed flood propagation indicated significantly lower losses than could be inferred from in-situ hydraulic properties, even allowing for subsurface heterogeneity. Possible causes are air entrapment, which could restrict infiltration rates, and the unknown effects of bed mobilisation and possible pore blockage by the heavy sediment loads transmitted under flood flow conditions.

A commonly observed effect is that, in the recession phase of the flow, deposition of a thin (1–2 mm) skin of fine sediment on the wadi bed occurs, which is sufficient to sustain flow over an unsaturated and transmissive wadi bed. Once the flow has ceased, this skin dries and breaks up so that the underlying alluvium is exposed for subsequent flow events. Crerar *et al.* (1988) observed from laboratory experiments that a thin continous silt layer was formed at low velocities. At higher velocities no such layer occurred as the bed surface was mobilised, but infiltration to the bed was still apparently inhibited. It was suggested that this could be due to clogging of the top layer of sand due to silt in the infiltrating water, or formation of a silt layer below the mobile upper part of the bed.

Further evidence for the heterogeneity of observed response comes from the observations of Hughes and Sami (1992) from a 39.6 km^2 semi-arid catchment in southern Africa. Soil moisture was monitored by neutron probe following two flow events. At some locations immediate response (monitored one day later) occurred throughout the profile; at others an immediate response near surface was followed by a delayed response at depth. Away from the inundated area, delayed response, assumed due to lateral subsurface transmission, occurred after 21 days.

The overall implication of the above observations is that it is not possible at present to extrapolate from in-situ point profile hydraulic properties to infer transmission losses from wadi channels. However, analysis of observed flood flows at different locations can allow quantification of losses, and studies by Walters (1990) and Jordan (1977), for example, provide evidence that the rate of loss is linearly related to the volume of surface discharge.

For southwest Saudi Arabia, the following relationships were defined:

$$LOSSL = 4.56 + 0.02216 \; UPSQ - 2034 \; SLOPE + 7.34$$
$$ANTEC(\text{s.e. } 4.15)$$
$$LOSSL = 3.75 \times 10^{-5} \; UPSQ^{0.821} \; SLOPE^{-0.865} \; ACWW^{0.497}$$
$$(\text{s.e. } 0.146 \text{ log units } (\pm 34\%))$$
$$LOSSL = 5.7 \times 10^{-5} \; UPSQ^{0.968} \; SLOPE^{-1.049}$$
$$(\text{s.e. } 0.184 \text{ loge units } (\pm 44\%))$$

where

LOSSL	= Transmission loss rate (1000 m^3 km^{-1})	(O.R. 1.08−87.9)
UPSQ	= Upstream hydrograph volume (1000 m^3)	(O.R. 69−3744)
SLOPE	= Slope of reach (mm^{-1})	(O.R. 0.001−0.011)
ANTEC	= Antecedent moisture index	(O.R. 0.10−1.00)
ACWW	= Active channel width (m)	(O.R. 25−231)

and

O.R. = Observed range

However, generalisation from limited experience can be misleading. Wheater *et al.* (1997) analysed transmission losses between two pairs of flow gauges on the Walnut Gulch catchment for a ten-year sequence and found that the simple linear model of transmission loss as proportional to upstream flow was inadequate. Considering the relationship

$$V_x = V_0(1 - \alpha)^x$$

where V_x is flow volume (m^3) at distance x downstream of flow volume V_0 and α represents the proportion of flow lost per unit distance, then α was found to decrease with discharge volume:

$$\alpha = 118.8 \; (V_0)^{-0.71}$$

The events examined had a maximum value of average transmission loss of 4076 m^3 km^{-1} in comparison with the estimate of Lane *et al.* (1971) of 4800–6700 m^3 km^{-1} as an upper limit of available alluvium storage.

The role of available storage was also discussed by Telvari *et al.* (1998), with reference to the Fowler's Gap catchment in Australia. Runoff plots were used to estimate runoff production as overland flow for a 4 km^2 basin. It was inferred that 7000 m^3 of overland flow becomes transmission loss, and that once this alluvial storage is satisfied, approximately two-thirds of overland flow is transmitted downstream.

A similar concept was developed by Andersen *et al.* (1998) at larger scale for the sand rivers of Botswana, which have alluvial beds of 20–200 m width and 2–20 m depth. Detailed observations of water table response showed that a single major event after a dry period of seven weeks was sufficient to fully satisfy available alluvial storage (the river bed reached full saturation within 10 hours). No significant drawdown occurred between subsequent events and significant resource potential remained throughout the dry season. It was suggested that two sources of transmission loss could be occurring: direct losses to the bed, limited by available storage; and losses through the banks during flood events.

It can be concluded that transmission loss is complex, that where deep unsaturated alluvial deposits exist the simple linear model as developed by Jordan (1977) and implicit in the results of Walters (1990) may be applicable, but that where alluvial storage is limited, this must be taken into account.

2.2.4 Groundwater recharge

While the estimation of groundwater recharge is essential to determine the sustainable yield of a groundwater resource, it is one of the most difficult hydrological fluxes to quantify. This is particularly so in arid and semi-arid areas, where values are low, and in any water-balance calculation groundwater recharge is likely to be lost in the uncertainty of the dominant inputs and outputs (i.e. precipitation and evaporation, together with another minor component of surface runoff).

It can be seen from the preceding discussion that groundwater recharge in arid and semi-arid areas is dependent on a complex set of spatial–temporal hydrological interactions that will be dependent on the characteristics of the local climate, the land surface properties that determine the balance between infiltration and overland flow, and the subsurface characteristics. Precipitation events in arid areas are generally infrequent but can be extremely intense. Where convective rainfall systems dominate, precipitation is highly localised in space and time (e.g. the monsoon rainfall of Walnut Gulch, Arizona), but in other climates more widespread winter rainfall may be important, and may include snow. Precipitation may infiltrate directly into the ground surface or, if overland flow is generated, be focused as surface runoff in a flowing channel. We have seen that for arid climates and convective rainfall, extensive overland flow can occur, enhanced by reduced surface soil permeabilities as a result of raindrop impact. The relationship between surface infiltration and groundwater recharge will depend on the subsurface properties, antecedent conditions, and the duration and intensity of the precipitation.

GROUNDWATER RECHARGE FROM EPHEMERAL CHANNEL FLOWS

The relationship between wadi flow transmission losses and groundwater recharge will depend on the underlying geology. The effect of lenses of reduced permeability on the infiltration process has been discussed and illustrated above, but once infiltration has taken place the alluvium underlying the wadi bed is effective in minimising evaporation loss through capillary rise (the coarse structure of alluvial deposits minimises capillary effects). Thus Hellwig (1973), for example, found that dropping the water table below 60 cm in sand with a mean diameter of 0.53 mm effectively prevented evaporation losses, and Sorey and Matlock (1969) reported that measured evaporation rates from streambed sand were lower than those reported for irrigated soils.

Parrisopoulos and Wheater (1991) combined two-dimensional simulation of unsaturated wadi bed response with Deardorff's (1977) empirical model of bare soil evaporation to show that evaporation losses were not in general significant for the water-balance or water table response in short-term simulation (i.e. for periods up to 10 days). However, the influence of vapour diffusion was not explicitly represented, and long-term losses are not well understood. Andersen et al. (1998) show that losses are high when the alluvial aquifer is fully saturated, but are small once the water table drops below the surface.

Sorman and Abdulrazzak (1993) provide an analysis of groundwater rise due to transmission loss for an experimental reach in Wadi Tabalah, southwest Saudi Arabia, and estimate that on average 75% of bed infiltration reaches the water table. There is in general little information available to relate flood transmission loss to groundwater recharge, however. The differences between the two are expected to be small, but will depend on residual moisture stored in the unsaturated zone and its subsequent drying characteristics. But if water tables approach the surface, relatively large evaporation losses may occur.

Again, it is tempting to draw over-general conclusions from limited data. In the study of the sand rivers of Botswana referred to above, it was expected that recharge of the alluvial river beds would involve complex unsaturated zone response. In fact, observations showed that the first flood of the wet season was sufficient to fully recharge the alluvial river bed aquifer. This storage was topped up in subsequent floods, and depleted by evaporation when the water table was near the surface, but in many sections sufficient water remained throughout the dry season to provide adequate sustainable water supplies for rural villages. And as noted above, Wheater et al. (1997) showed for Walnut Gulch and Telvari et al. (1998) for Fraser's Gap that limited river bed storage affected transmission loss. It is evident that surface-water/groundwater interactions depend strongly on the local characteristics of the underlying alluvium and the extent of their connection to, or isolation from, other aquifer systems.

Recent work at Walnut Gulch (Goodrich et al., 2004) has investigated ephemeral channel recharge using a range of experimental methods, combined with modelling. These consisted of a reach water-balance method, including estimates of near channel evapotranspiration losses, geochemical methods, analysis of changes in groundwater levels and microgravity measurements, and unsaturated zone flow and temperature analyses. The conclusions were that ephemeral channel losses were significant as an input to the underlying regional aquifer, and that the range of methods for recharge estimation agreed within a factor of 3 (reach water-balance methods giving the higher estimates).

SPATIAL VARIABILITY OF GROUNDWATER RECHARGE

The above discussion of groundwater recharge from ephemeral channels highlighted the role of subsurface properties in determining the relationship between surface infiltration and groundwater recharge. The same issues apply to the relationships between infiltration and recharge away from the active channel system, but data to quantify the process response associated with distributed recharge are commonly lacking.

An important requirement for recharge estimation has arisen in connection with the proposal for a repository for high-level nuclear waste at Yucca Mountain, Nevada, and a major effort has been made to investigate the hydrological response of the deep unsaturated zone which would house the repository. Flint et al. (2002) propose that soil depth is an important control on infiltration. They argue that groundwater recharge (or net infiltration) will occur when the storage capacity of the soil or rock is exceeded. Hence, where thin soils are underlain by fractured bedrock, a relatively small amount of infiltration is required to saturate the soil and generate 'net infiltration', which will occur at a rate dependent on the bedrock (fracture) permeability. Deeper soils have greater storage and provide greater potential to store infiltration that will subsequently be lost as evaporation. This conceptual understanding was encapsulated in a distributed hydrological model, incorporating soil depth variability and the potential for redistribution through overland flow. The results indicate extreme variability in space and time, with watershed modelling giving a range from zero to several hundred millimetres per year, depending on spatial location. The high values arise due to flow focusing in ephemeral channels and subsequent channel bed infiltration. The need to characterise in detail the response of a deep unsaturated zone led to the application at Yucca Mountain of a wide range of alternative methods for recharge estimation. Hence, Flint *et al.* (2002) also review results from methods including analysis of physical data from unsaturated zone profiles of moisture and heat, and the use of environmental tracers. These methods operate at a range of spatial scales, and the results support the modelling conclusions that great spatial variability of recharge can be expected at the site scale, at least for comparable systems of fractured bedrock.

The Yucca Mountain work illustrates that high spatial variability occurs at the relatively small spatial scale of an individual site. More commonly, a broader scale of assessment is needed, and broader scale features become important. For example, Wilson and Guan (2004) introduce the term 'mountain-front recharge' (MFR) to describe the important role for many aquifer systems in arid or semi-arid areas of the contribution from mountains to the recharge of aquifers in adjacent basins. They argue that mountains have higher precipitation, cooler temperatures (and hence less evaporation) and thin soils, all contributing to greater runoff and recharge, and that in many arid and semi-arid areas this will dominate over the relatively small contribution to recharge from direct infiltration of precipitation from the adjacent arid areas.

It is clear from the above discussion that appropriate representation of spatial variability of precipitation and the subsequent hydrological response is required for the estimation of recharge at catchment or aquifer scale, and that even at local scale these effects are important.

2.3 HYDROLOGICAL MODELLING AND THE REPRESENTATION OF RAINFALL

The preceding discussion illustrates some of the particular characteristics of arid areas which place special requirements on hydrological modelling, for example for flood, water resources or groundwater recharge estimation. One evident area of difficulty is rainfall, especially where convective storms are an important influence. The work of Michaud and Sorooshian (1994) demonstrated the sensitivity of flood peak simulation to the spatial resolution of rainfall input. This obviously has disturbing implications for flood modelling, particularly where data

availability is limited to conventional raingauge densities. Indeed, it appears highly unlikely that suitable raingauge densities will ever be practicable for routine monitoring. However, the availability of 2 km resolution radar data in the USA can provide adequate information and radar could be installed elsewhere for particular applications. Morin *et al.* (1995) report results from a radar located at Ben-Gurion airport in Israel, for example. Where convective rainfall predominates, rainfall variability is extreme and raises specific issues of data and modelling. However, in general, as seen from the discussion of mountain front recharge, the spatial distribution of precipitation is important.

One way forward for the problems of convective rainfall is to develop an understanding of the properties of spatial rainfall based on high-density experimental networks and/or radar data, and represent those properties within a spatial rainfall model for more general application. It is likely that this would have to be done within a stochastic modelling framework in which equally likely realisations of spatial rainfall are produced, possibly conditioned by sparse observations.

Some simple empirical first steps in this direction were taken by Wheater *et al.* (1991a,b) for southwest Saudi Arabia and Wheater *et al.* (1995) for Oman. In the Saudi Arabian studies, as noted earlier, raingauge data was available at approximately 10 km spacing and spatial correlation was low; hence, a multivariate model was developed, assuming independence of raingauge rainfall. Based on observed distributions, seasonally dependent catchment rainday occurrence was simulated, dependent on whether the preceding day was wet or dry. The number of gauges experiencing rainfall was then sampled, and the locations selected based on observed occurrences (this allowed for increased frequency of raindays with increased elevation). Finally, start-times, durations and hourly intensities were generated. Model performance was compared with observations. Rainfall from random selections of raingauges was well reproduced, but when clusters of adjacent gauges were evaluated a degree of spatial organisation of occurrence was observed but not simulated. It was evident that a weak degree of correlation was present, which should not be neglected. Hence, in extension of this approach to Oman (Wheater *et al.*, 1995), observed spatial distributions were sampled, with satisfactory results.

However, this multivariate approach suffers from limitations of raingauge density, and in general a model in continuous space (and continuous time) is desirable. A family of stochastic rainfall models of point rainfall was proposed by Rodriguez-Iturbe *et al.* (1987, 1988) and applied to UK rainfall by Onof and Wheater (1993, 1994). The basic concept is that a Poisson process is used to generate the arrival of storms. Associated with a storm is the arrival of raincells of uniform intensity for a given duration (sampled from specified distributions). The overlapping of these rectangular pulse cells generates the storm

Table 2.3. *Performance of the Bartlett–Lewis rectangular pulse model in representing July rainfall at gauge 44, Walnut Gulch.*

	Mean	Var	ACF1	ACF2	ACF3	Pwet	Mint	Mno	Mdur
Model	0.103	1.082	0.193	0.048	0.026	0.032	51.17	14.34	1.68
Data	0.100	0.968	0.174	0.040	0.036	0.042	53.71	13.23	2.38

intensity profile in time. These models were shown to have generally good performance for the UK in reproducing rainfall properties at different time-scales (from hourly upwards) and extreme values.

Cox and Isham (1988) extended this concept to a model in space and time, whereby the raincells are circular and arrive in space within a storm region. As before, the overlapping of cells produces a complex rainfall intensity profile, now in space as well as time. This model has been developed further by Northrop (1998) to include elliptical cells and storms and is being applied to UK rainfall (Northrop *et al.* 1999).

Work by Samuel (1999) explored the capability of these models to reproduce the convective rainfall of Walnut Gulch. In modelling point rainfall, the Bartlett–Lewis rectangular pulse model was generally slightly superior to other model variants tested. Table 2.3 shows representative performance of the model in comparing the hourly statistics from 500 realisations of July rainfall in comparison with 35 years from one of the Walnut Gulch gauges (gauge 44), where: Mean is the mean hourly rainfall (mm); Var its variance; ACF1, ACF2 and ACF3 the autocorrelations for lags 1, 2, 3; Pwet the proportion of wet intervals; Mint the mean storm inter-arrival time (h); Mno the mean number of storms per month; and Mdur the mean storm duration (h). This performance is generally encouraging (although the mean storm duration is underestimated), and extreme value performance is excellent.

Work with the spatial–temporal model was only taken to a preliminary stage, but Figure 2.5 shows a comparison of observed spatial coverage of rainfall for 25 years of July data from 81 gauges (for different values of the standard deviation of cell radius) and Figure 2.6 the corresponding fit for temporal lag-0 spatial correlation. Again, the results are encouraging, and there is promise with this approach to address the significant problems of spatial representation of convective rainfall for hydrological modelling.

Where spatial rainfall variability is less extreme and coarser time-scale modelling is appropriate (e.g. daily rainfall), recent work has developed a set of stochastic rainfall modelling tools that can be readily applied to represent rainfall, including effects of location (e.g. topographic effects or rainshadow) and climate variability and change (e.g. Chandler and Wheater, 2002; Yang *et al.*, 2005). These use generalised linear models (GLMs) to simulate the occurrence of a rainday, and then the conditional distribution of daily rainfall depths at selected locations over an

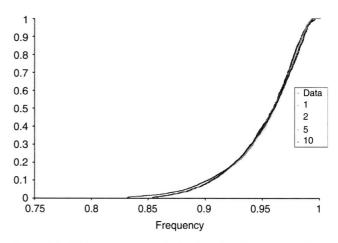

Figure 2.5 (Plate 1). Frequency distribution of spatial coverage of Walnut Gulch rainfall. Observed vs. alternative simulations.

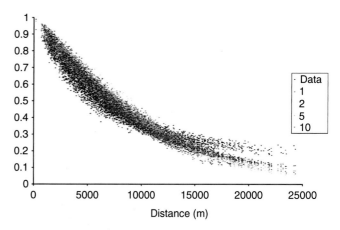

Figure 2.6 (Plate 2). Spatial correlation of Walnut Gulch rainfall. Observed vs. alternative simulations.

area. Although initially developed and evaluated for humid temperate climates, work is currently underway to evaluate their use for drier climates, with application to Iran (Mirshahi *et al.*, 2008) and Botswana (Kenabatho *et al.*, 2008).

2.4 INTEGRATED MODELLING FOR GROUNDWATER RESOURCE EVALUATION

Appropriate strategies for water resource development must recognise the essential physical characteristics of the hydrological

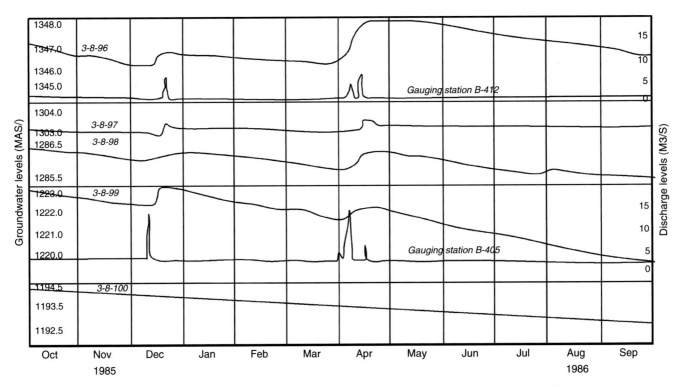

Figure 2.7. Longitudinal sequence of wadi alluvium well hydrographs and associated surface flows, Wadi Tabalah, 1985/86.

processes. As noted above, groundwater is a resource particularly well suited to arid regions. Subsurface storage minimises evaporation loss and can provide long-term yields from infrequent recharge events. The recharge of alluvial groundwater systems by ephemeral flows can provide an appropriate resource, and this has been widely recognised by traditional development, such as the 'afalaj' of Oman and elsewhere. There may, however, be opportunities for augmenting recharge and more effectively managing these groundwater systems. In any case, it is essential to quantify the sustainable yield of such systems, for appropriate resource development.

It has been seen that observations of surface flow are ambiguous, due to upstream tansmission loss, and do not define the available resource. Similarly, observed groundwater response does not necessarily indicate upstream recharge. Figure 2.7 presents a series of groundwater responses from 1985/86 for Wadi Tabalah which shows a downstream sequence of wells 3-B-96, -97, -98, -99 and -100 and associated surface-water discharges. It can be seen that there is little evidence of the upstream recharge at the downstream monitoring point.

In addition, records of surface flows and groundwater levels, coupled with ill-defined histories of abstraction, are generally insufficient to define long-term variability of the available resource.

To capture the variability of rainfall and the effects of transmission loss on surface flows, a distributed approach is necessary. If groundwater is to be included, integrated modelling of surface water and groundwater is needed. Examples of distributed surface-water models include KINEROS (Wheater and Bell, 1983; Michaud and Sorooshian, 1994; Al-Qurashi et al., 2008), the model of Sharma (1997, 1998) and a distributed model, INFIL2.0, developed by the USGS to simulate net infiltration at Yucca Mountain (Flint et al., 2000).

A distributed approach to the integrated modelling of surface and groundwater response following Wheater et al. (1995) is illustrated in Figure 2.8. This schematic framework requires the characterisation of the spatial and temporal variability of rainfall, distributed infiltration, runoff generation and flow transmission losses, the ensuing groundwater recharge and groundwater response. This presents some technical difficulties, although the integration of surface and groundwater modelling allows maximum use to be made of available information, so that, for example, groundwater response can feed back information to constrain surface hydrological parameterisation. A distributed approach provides the only feasible method of exploring the internal response of a catchment to management options.

This integrated modelling approach was developed for Wadi Ghulaji, Sultanate of Oman, to evaluate options for groundwater recharge management (Wheater et al., 1995). The catchment, of area 758 km[2], drains the southern slopes of Jebal Hajar in the Sharqiyah region of northern Oman. Proposals to be evaluated included recharge dams to attenuate surface flows and provide managed groundwater recharge in key locations. The modelling framework involved the coupling of a distributed rainfall model, a distributed water-balance model (incorporating rainfall-runoff processes, soil infiltration and wadi flow transmission losses), and a distributed groundwater model (Figure 2.9).

The representation of rainfall spatial variability presents technical difficulties, since data are limited. Detailed analysis was undertaken of 19 raingauges in the Sharqiyah region, and of 6 raingauges in the catchment itself. A stochastic multivariate temporal–spatial model was devised for daily rainfall, a modified version of a scheme orgininally developed by Wheater *et al.* (1991a,b).

The occurrence of catchment rainfall was determined according to a seasonally variable first-order Markov process, conditioned on rainfall occurrence from the previous day. The number and locations of active raingauges and the gauge depths were derived by random sampling from observed distributions.

The distributed water-balance model represents the catchment as a network of two-dimensional plane and linear channel elements. Runoff and infiltration from the planes was simulated using the US Soil Conservation Service (SCS) approach. Wadi flows incorporate a linear transmission-loss algorithm based on work by Jordan (1977) and Walters (1990). Distributed calibration parameters are shown in Figure 2.10.

Finally, a groundwater model was developed based on a detailed hydrogeological investigation which led to a multilayer representation of uncemented gravels, weakly/strongly cemented gravels and strongly cemented/fissured gravel/bedrock, using MODFLOW.

The model was calibrated to the limited flow data available (a single event) (Table 2.4), and was able to reproduce the distribution of runoff and groundwater recharge within the catchment through a rational association on loss parameters with topography, geology and wadi characteristics. Extended synthetic data sequences were then run to investigate catchment water balances under scenarios of different runoff exceedance probabilities (20%, 50%, 80%), as in Table 2.4, and to investigate management options.

This example, although based on limited data, is not untypical of the requirements to evaluate water resource management options in practice, and the methodology can be seen as providing a generic basis for the assessment of management options. An integrated modelling approach provides a powerful basis for assimilating available hard and soft surface and groundwater data in the assessment of recharge, as well as providing a management tool to explore effects of climate variability and management options.

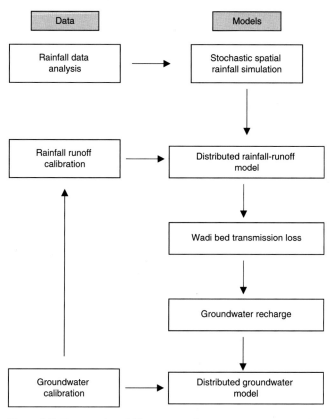

Figure 2.8. Integrated modelling strategy for water resource evaluation.

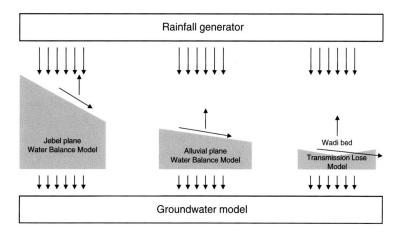

Figure 2.9. Schematic of the distributed water resource model.

Table 2.4. *Annual catchment water balance, simulated scenarios.*

Scenario	Rainfall	Evaporation	Groundwater recharge	Runoff	% Runoff
Wet	88	0.372	12.8	4.0	4.6
Average	33	0.141	11.2	3.5	4.0
Dry	25	0.106	5.5	1.7	3.2

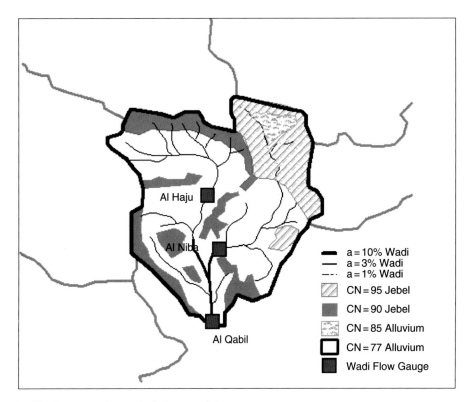

Figure 2.10. Distributed calibration parameters, water-balance model.

2.5 CONCLUSIONS

This chapter has attempted to illustrate the hydrological characteristics of arid and semi-arid areas, which are complex and generally poorly understood. For many hydrological applications, including the estimation of groundwater recharge and surface/groundwater interactions, these characteristics present severe problems for conventional methods of analysis. Much high-quality experimental research is needed to develop knowledge of spatial rainfall, runoff processes, infiltration and groundwater recharge, and to understand the role of vegetation and of climate variability on runoff and recharge processes. Recent data have provided new insights, and there is a need to build on these to develop appropriate methods for water resource evaluation and management, and in turn, to define data needs and research priorities.

Groundwater recharge is one of the most difficult fluxes to define, particularly in arid and semi-arid areas. However, reliable quantification is essential to maximise the potential of groundwater resources, define long-term sustainable yields and protect traditional sources. This requires an appropriate conceptual understanding of the important processes and their spatial variability, and the assimilation of all relevant data, including surface-water and groundwater information. It is argued that distributed modelling of the integrated surface-water/groundwater system is a valuable if not essential tool, and a generic framework to achieve this has been defined and a simple application example presented. However, such modelling requires that the dominant processes are characterised, including rainfall, rainfall-runoff processes, infiltration and groundwater recharge, and the detailed hydrogeological response of what are often complex groundwater systems.

REFERENCES

Allison, G. B. and Hughes, M. W. (1978) The use of environmental chloride and tritium to estimate total recharge to an unconfined aquifer. *Aust. J. Soil Res.* **16**, 181–195.

Al-Qurashi, A., McIntyre, N., Wheater, H. and Unkrich, C. (2008) Application of the Kineros2 rainfall-runoff model to an arid catchment in Oman. *J. Hydrol.* **355**, 91–105.

Andersen, N. J., Wheater, H. S., Timmis, A. J. H. and Gaongalelwe, D. (1998). Sustainable development of alluvial groundwater in sand rivers of Botswana. In *Sustainability of Water Resources under Increasing Uncertainty*, IAHS Publ. **240**, 367–376.

Bromley, J., Edmunds, W. M., Fellman E. et al. (1997) Estimation of rainfall inputs and direct recharge to the deep unsaturated zone of southern Niger using the chloride profile method. *J. Hydrol.* **188–189**, 139–154.

Chandler, R. E. and Wheater, H. S. (2002) Analysis of rainfall variability using generalized linear models: a case study from the West of Ireland. *Water Resour. Res.* **38**(10), 1192, 10–1–10–11.

Cordery, I., Pilgrim, D. H. and Doran, D. G. (1983) Some hydrological characteristics of arid western New South Wales. The Institution of Engineers, Australia, Hydrology and Water Resources Symp., Nov. 1983.

Cox, D. R. and Isham, V.(1988) A simple spatial–temporal model of rainfall, *Proc. Roy. Soc.* **A415**, 317–328.

Crerar, S., Fry, R. G., Slater, P. M., van Langenhove, G. and Wheeler, D. (1988) An unexpected factor affecting recharge from ephemeral river flows in SWA/Namibia. In *Estimation of Natural Groundwater Recharge*, ed. I. Simmers, 11–28. D. Reidel.

Deardorff, J. W. (1977) A parameterization of ground-surface moisture content for use in atmospheric prediction models. *J. Applied. Meteorol.* **16**, 1182–1185.

Desconnets, J. C., Taupin, J. D., Lebel, T. and Leduc, C. (1997) Hydrology of the HAPEX-Sahel Central Super-Site: surface water drainage and aquifer recharge through the pool systems. *J. Hydrol.* **188**–189, 155–178.

Flint, A. L., Flint, L. E., Kwicklis, E. M., Fabryka-Martin, J. T. and Bodvarsson, G. S. (2002) Estimating recharge at Yucca Mountain, Nevada, USA: comparison of methods. *Hydrogeology Journal* **10**, 180–204.

Food and Agriculture Organization (1981) Arid zone hydrology for agricultural development. *Irrig. Drain. Pap.* **37**. FAO.

Goodrich, D. C., Lane, L. J., Shillito, R. M. et al. (1997) Linearity of basin response as a function of scale in a semi-arid watershed. *Water Resour. Res.*, **33**(12), 2951–2965.

Goodrich, D. C., Williams, D. G., Unkrich, C. L. et al. (2004) Comparison of methods to estimate ephemeral channel recharge, Walnut Gulch, San Pedro River Basin, Arizona. In *Groundwater Recharge in a Desert Environment. The Southwestern United States*, ed. J. F. Hogan, F. M. Phillips and B. R. Scanlon, 77–99. Water Science and Application 9, American Geophysical Union.

Goutorbe, J. P., Dolman, A. J., Gash, J. H. C. et al. (eds) (1997) *HAPEX-Sahel*. Elsevier (reprinted from *J. Hydrol.* **188–189**, 1–4).

Hellwig, D. H. R. (1973) Evaporation of water from sand. 3. The loss of water into the atmosphere from a sandy river bed under arid climatic condtions. *J. Hydrol.* **18**, 305–316.

Hughes, D. A. (1995) Monthly rainfall-runoff models applied to arid and semiarid catchments for water resource estimation purposes. *Hydr. Sci. J.* **40**(6), 751–769.

Hughes, D. A. and Sami, K. (1992) Transmission losses to alluvium and associated moisture dynamics in a semiarid ephemeral channel system in Southern Africa. *Hydrological Processes*, **6**, 45–53.

Jordan, P. R. (1977) Streamflow transmission losses in Western Kansas. *Jnl of Hydraulics Division, ASCE*, **108**, HY8, 905–919.

Kenabatho, P. K., McIntyre, N. R. and Wheater, H. S. (2008) Application of generalised linear models for rainfall simulations in semi arid areas: a case study from the Upper Limpopo basin in north east Botswana. 10th BHS National Hydrology Symposium, Exeter, Sept. 2008.

Lane, L. J., Diskin, M. H. and Renard, K. G. (1971) Input–output relationships for an ephemeral stream channel system. *J. Hydrol.* **13**, 22–40.

Lebel, T., Taupin, J. D. and D'Amato, N. (1997) Rainfall monitoring during HAPEX-Sahel. 1. General rainfall conditions and climatology. *J. Hydrol.* **188–189**, 74–96.

Lebel, T. and Le Barbe, L. (1997) Rainfall monitoring during HAPEX-Sahel. 2. Point and areal estimation at the event and seasonal scales. *J. Hydrol.* **188–189**, 97–122.

Liu, B., Phillips, F., Hoines, S., Campbell, A. R. and Sharma, P. (1995) Water movement in desert soil traced by hydrogen and oxygen isotopes, chloride, and chlorine-36, southern Arizona. *J. Hydrol.* **168**, 91–110.

Macmillan, L. C. (1987) *Regional evaporation and soil moisture analysis for irrigation application in arid areas*. Univ. London MSc thesis, Dept of Civil Engineering, Imperial College.

McIntyre, N., Al-Qurashi, A. and Wheater, H. S. (2007) Regression analysis of rainfall-runoff data from an arid catchment in Oman. *Hydrological Sciences Journal* **52**(6), 1103–1118, Dec. 2007.

McMahon, T. A. (1979) Hydrological characteristics of arid zones. *Proceedings of a Symposium on the Hydrology of Areas of Low Precipitation, Canberra*. IAHS Publ. **128**, 105–123.

Michaud, J. D. and Sorooshian, S. (1994) Effect of rainfall-sampling errors on simulations of desert flash floods. *Water Resour. Res.* **30**(10), 2765–2775.

Mirshahi, B., Onof, C. J. and Wheater, H. S. (2008) Spatial–temporal daily rainfall simulation for a semi-arid area in Iran: a preliminary evaluation of generalised linear models. Proc. 10th BHS National Hydrology Symposium, Exeter, Sept. 2008.

Mocke, R. (1998) *Modelling the sand-rivers of Botswana – distributed modelling of runoff and groundwater recharge processes to assess the sustainability of rural water supplies*. Univ. London MSc thesis, Dept of Civil Engineering, Imperial College.

Morin, J. and Benyamini, Y. (1977) Rainfall infiltration into bare soils. *Water. Resour. Res.* **13**(5), 813–817.

Morin, J., Rosenfeld, D. and Amitai, E. (1995) Radar rain field evaluation and possible use of its high temporal and spatial resolution for hydrological purposes. *J. Hydrol.* **172**, 275–292.

Nemec, J. and Rodier, J. A. (1979) Streamflow characteristics in areas of low precipitation. *Proceedings of a Symposium on the Hydrology of Areas of Low Precipitation, Canberra*. IAHS Publ. **128**, 125–140.

Northrop, P. J. (1998) A clustered spatial–temporal model of rainfall. *Proc. Roy. Soc.* **A454**, 1875–1888.

Northrop, P. J., Chandler, R. E., Isham, V. S., Onof, C. and Wheater, H. S. (1999) Spatial–temporal stochastic rainfall modelling for hydrological design. In *Hydrological Extremes: Understanding, Predicting, Mitigating*, ed. L. Gottschalk, J.-C. Olivry, D. Reed and D. Rosbjerg. IAHS Publ. **255**, 225–235.

Onof, C. and Wheater, H. S. (1993) Modelling of British rainfall using a random parameter Bartlett-Lewis rectangular pulse model. *J. Hydrol.* **149**, 67–95.

Onof, C. and Wheater, H. S. (1994) Improvements of the modelling of British rainfall using a modified random parameter Bartlett-Lewis rectangular pulse model. *J. Hydrol.* **157**, 177–195.

Osborn, H. B. and Reynolds, W. N. (1963) Convective storm patterns in the Southwestern United States. *Bull. IASH* **8**(3), 71–83.

Osborn, H. B., Lane, L. J. and Hundley, J. F. (1972) Optimum gaging of thunderstorm rainfall in southeastern Arizona, *Water Resour. Res.* **8**(1), 259–265.

Osborn, H. B., Renard, K. G. and Simanton, J. R. (1979) Dense networks to measure convective rainfalls in the Southwestern United States. *Water Resour. Res.* **15**(6), 1701–1711.

Parissopoulos, G. A. and Wheater, H. S. (1990) Numerical study of the effects of layers on unsaturated-saturated two-dimensional flow. *Water Resources Mgmt.* **4**, 97–122.

Parissopoulos, G. A., and Wheater, H. S. (1991) Effects of evaporation on groundwater recharge from ephemeral flows. In *Advances in Water Resources Technology*, ed. G. Tsakiris and A. A. Balkema, 235–245, Balkema.

Parrisopoulos, G. A., and Wheater, H. S. (1992a) Experimental and numerical infiltration studies in a wadi stream-bed. *J. Hydr. Sci.* **37**, 27–37.

Parissopoulos, G. A., and Wheater, H. S. (1992b) Effects of hysteresis on groundwater recharge from ephemeral flows. *Water Resour. Res.* **28**(11), 3055–3061.

Pilgrim, D. H., Chapman, T. G. and Doran, D. G. (1988) Problems of rainfall-runoff modelling in arid and semi-arid regions. *Hydrol. Sci. J.* **33**(4), 379–400.

Rodriguez-Iturbe, I., Cox, D. R. and Isham, V. (1987) Some models for rainfall based on stochastic point processes. *Proc. Roy. Soc.* **A410**, 269–288.

Rodriguez-Iturbe, I., Cox, D. R. and Isham, V. (1988) A point process model for rainfall: further developments. *Proc. Roy. Soc.* **A417**, 283–298.

Samuel, C. R. (1999) *Stochastic rainfall modelling of convective storms in Walnut Gulch, Arizona*. Univ. London PhD Thesis.

Saudi Arabian Dames and Moore (1988) *Representative Basins Study*. Final Report to Ministry of Agriculture and Water, Riyadh, 84 vols.

Sharma, K. D. (1997) Integrated and sustainable development of water-resources of the Luni basin in the Indian arid zone. In *Sustainability of Water Resources under Increasing Uncertainty*. IAHS Publ. **240**, 385–393.

Sharma, K. D. (1998) Resource assessment and holistic management of the Luni River Basin in the Indian desert. In *Hydrology in a Changing Environment*, Vol 2, ed. H. Wheater and C. Kirby, 387–395, Wiley.

Sorey, M. L. and Matlock, W. G. (1969) Evaporation from an ephemeral streambed. *J. Hydraul. Div. Am. Soc. Civ. Eng.* **95**, 423–438.

Sorman, A. U. and Abdulrazzak, M. J. (1993) Infiltration–recharge through wadi beds in arid regions. *Hydr. Sci. Jnl.* **38**(3), 173–186.

Telvari, A., Cordery, I. and Pilgrim, D. H. (1998) Relations between transmission losses and bed alluvium in an Australian arid zone stream. In *Hydrology in a Changing Environment*, Vol **2**, ed. H. Wheater and C. Kirby, 361–366, Wiley.

UNDP (1992) *Surface Water Resources*. Final report to the Government of the Republic of Yemen High Water Council UNDP/DESD PROJECT YEM/88/001 Vol III, Jun. 1992.

Walters, M. O. (1990) Transmission losses in arid region. *Jnl of Hydraulic Engineering* **116**(1), 127–138.

Wheater, H. S. and Bell, N. C. (1983) Northern Oman flood study. *Proc. Instn. Civ. Engrs. Part 2*, **75**, 453–473.

Wheater, H. S., and Brown, R. P. C. (1989) Limitations of design hydrographs in arid areas – an illustration from southwest Saudi Arabia. *Proc. 2nd Natl. BHS Symp. (1989)*, 3.49–3.56.

Wheater, H. S., Larentis, P. and Hamilton, G. S. (1989) Design rainfall characteristics for southwest Saudi Arabia. *Proc. Inst. Civ. Eng., Part 2*, **87**, 517–538.

Wheater, H. S., Butler, A. P., Stewart, E. J. and Hamilton, G. S. (1991a) A multivariate spatial–temporal model of rainfall in S.W. Saudi Arabia. I. Data characteristics and model formulation. *J. Hydrol.* **125**, 175–199.

Wheater, H. S., Jolley, T. J. and Peach, D. (1995) A water resources simulation model for groundwater recharge studies: an application to Wadi Ghulaji, Sultanate of Oman. In *Proc. Intl. Conf. on Water Resources Management in Arid Countries (Muscat)*, 502–510.

Wheater, H. S., Woods Ballard, B. and Jolley, T. J. (1997) An integrated model of arid zone water resources: evaluation of rainfall-runoff simulation performance. In *Sustainability of Water Resources under Increasing Uncertainty*, IAHS Publ. **240**, 395–405.

Wilson, J. L. and Guan, H. (2004) Mountain-block hydrology and Mountain-front recharge. In *Groundwater Recharge in a Desert Environment. The Southwestern United States*, ed. J. F. Hogan, F. M. Phillips and B. R. Scanlon, 113–137. Water Science and Application 9, American Geophysical Union.

Yang, C., Chandler, R. E., Isham, V. S. and Wheater, H. S. (2005) Spatial–temporal rainfall simulation using generalized linear models. *Water Resources Research*, **41**, W11415, doi 10.1029/2004 WR003739.

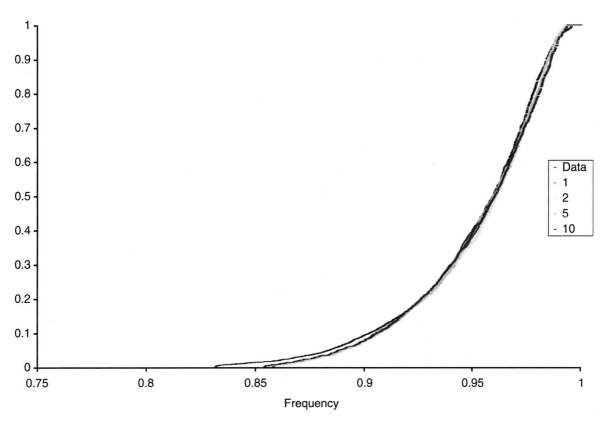

Plate 1 (Figure 2.5). Frequency distribution of spatial coverage of Walnut Gulch rainfall. Observed vs. alternative simulations.

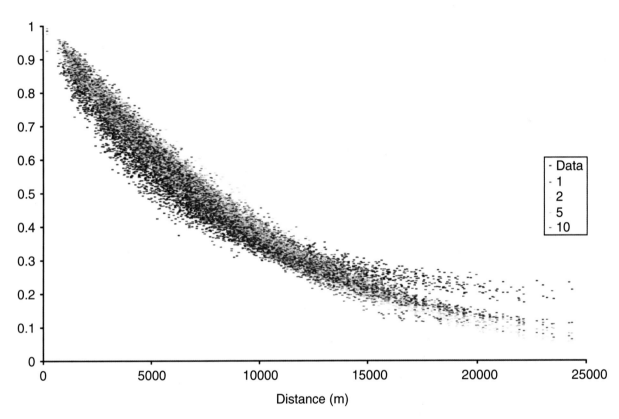

Plate 2 (Figure 2.6). Spatial correlation of Walnut Gulch rainfall. Observed vs. alternative simulations.

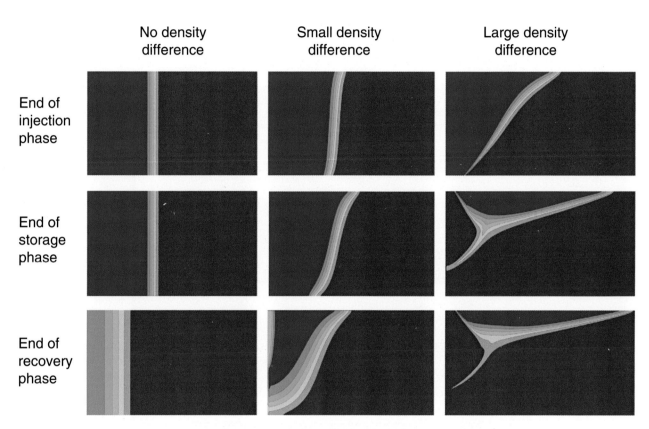

Plate 3 (Figure 7.12). An illustrative comparison between FEFLOW numerical model results for the cases of no density, small density and large density contrasts. Results are given for: end of injection phase, end of storage phase and end of recovery phase. Each image is symmetrical about its left-hand boundary, which represents the injection/extraction well where freshwater is injected into the initially more saline aquifer. The effects of variable density flow are clearly visible. Increasing the density contrast results in deviation from the standard cylindrical 'bubble' most, as seen in the no density difference case.

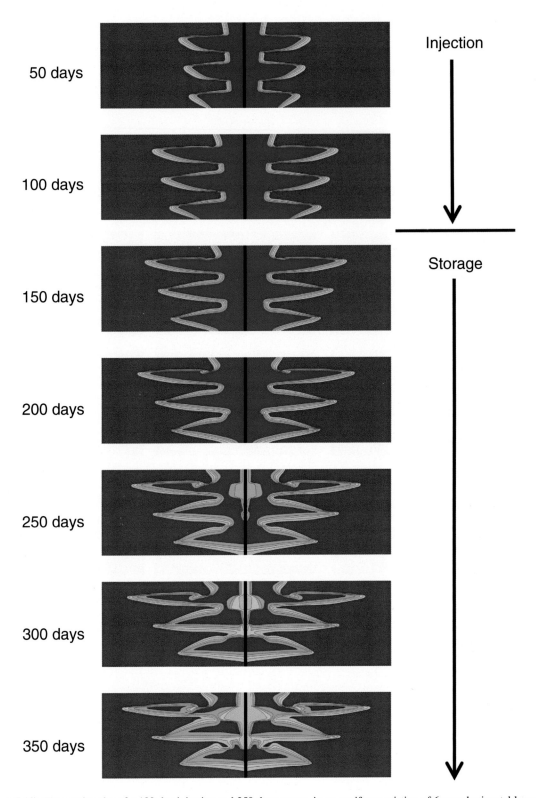

Plate 4 (Figure 7.13). Time series plot of a 100-day injection and 250-day storage, in an aquifer consisting of 6 even horizontal layers, with the hydraulic conductivity varying by a ratio of 10 between high and low K layers. Note the fingering processes.

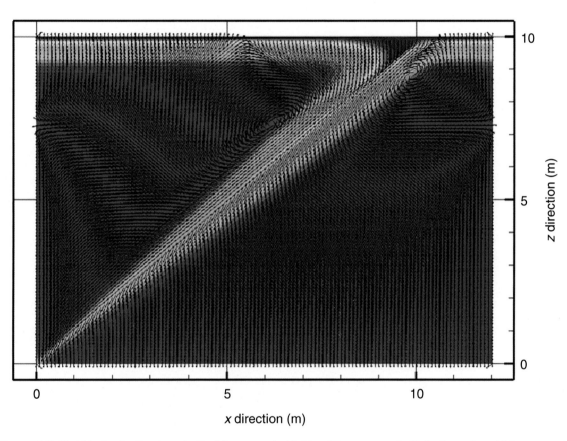

Plate 5 (Figure 7.14). Variable density flow in an inclined fracture embedded in a 2D porous matrix. Red colors refer to high concentration and blue colors refer to low concentration. Arrows represent groundwater flow velocities.

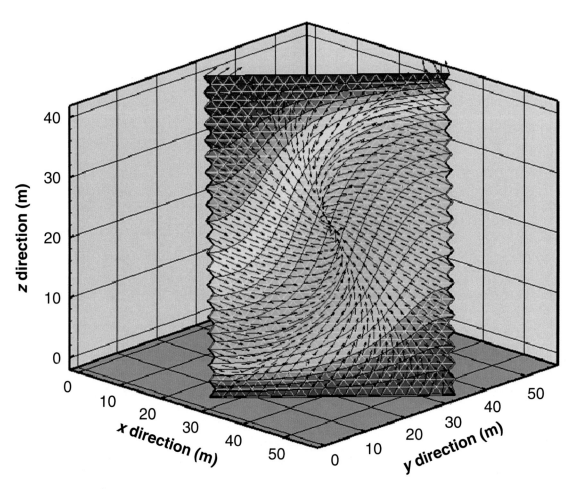

Plate 6 (Figure 7.15). Preliminary results of variable density flow simulations in an inclined non-planar fracture embedded in a 3D porous matrix. The fracture is discretised by 9 planar triangles +(triangulation) and the porous matrix cube is of side length 10m.

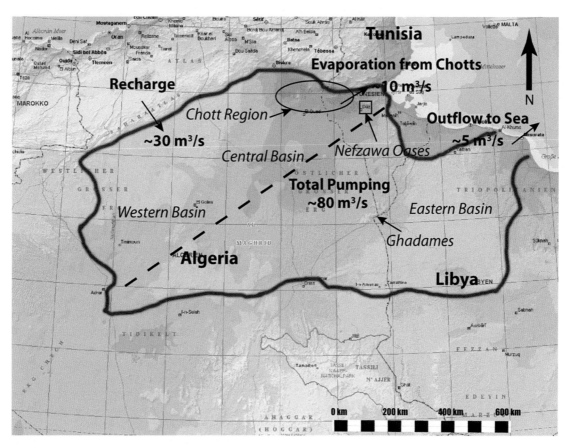

Plate 7 (Figure 8.1). Map of northern Africa showing the outline of the NWSAS as well as the estimated water balance for the year 2000 (Siegfried, 2004).

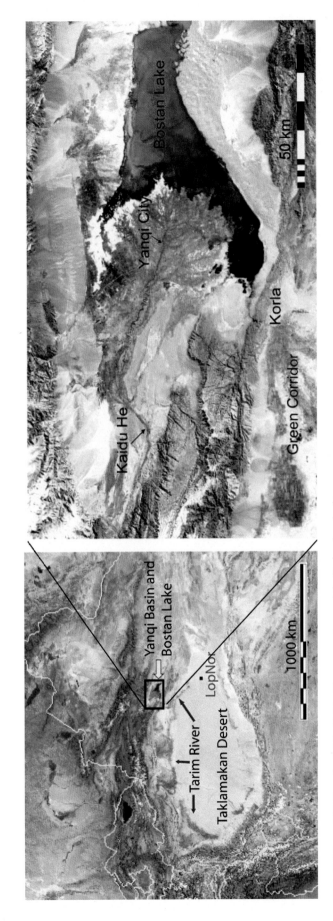

Plate 8 (Figure 8.3). Location of the Kaidu–Kongque system and the Tarim river basin.

3 Conceptual models for recharge sequences in arid and semi-arid regions using isotopic and geochemical methods

W. M. Edmunds

3.1 INTRODUCTION

Groundwater quality issues now assume major importance in semi-arid and arid zones, since scarce reserves of groundwater (both renewable and non-renewable) are under threat due to accelerated development as well as a range of direct human impacts. In earlier times, settlement and human migration in Africa and the Middle East regions were strictly controlled by the locations and access to fresh water mainly close to the few major perennial rivers such as the Nile, or to springs and oases, representing key discharge points of large aquifers that had been replenished during wetter periods of the Holocene and the Pleistocene. This is a recurring theme in writings of all the early civilisations where water is valued and revered (Issar, 1990). The memory of the Holocene sea-level rise and pluvial periods is indicated in the story of the Flood (Genesis 7:10) and groundwater quality is referred to in the book of Exodus (15:20). More recent climate change may also be referred to in the Koran (Sura 34:16).

In present times, groundwater forms the primary source of drinking water in arid and semi-arid regions, since river flows are unreliable and large freshwater lakes are either ephemeral (e.g. Lake Chad) or no longer exist. The rules of use and exploitation have changed dramatically over the past few decades by the introduction of advanced drilling technology for groundwater (often available alongside oil exploitation), as well as the introduction of mechanised pumps. This has raised expectations in a generation or so of the availability of plentiful groundwater, yet in practice falling water levels testify to probable over-development and inadequate scientific understanding of the resource, or to failure of management to act on the scientific evidence available. In addition to the issues related to quantity, quality issues exacerbate the situation. This is because the natural groundwater regime established over long time-scales has developed chemical (and age) stratification in response to recharge over a range of climatic regimes and geological controls. Borehole drilling cuts through the natural quality layering, and abstraction may lead to deterioration in water quality with time as water either is drawn from lower transmissivity strata, or is drawn down from the near surface where saline waters are commonplace.

Investigations of water quality are therefore needed alongside and even in advance of widespread exploitation. The application of hydrogeochemical techniques is needed to fully understand the origins of groundwater, the timing of its recharge and whether or not modern recharge is occurring at all. The present-day flow regime may differ from that at the time of recharge due to major changes in climate from the original wetter recharge periods; this may affect modelling of groundwater in semi-arid and arid regions. Geochemical investigation may help to create and to validate suitable models.

Study is needed of the extent of natural layering and zonation as well as chemical changes taking place along flow lines due to water–rock interaction (predicting, for example, if harmful concentrations of certain elements may be present naturally). The superimposed impacts of human activity and the ability of the aquifer system to act as a buffer against surface pollution can then be determined against natural baselines.

Typical landscape elements of arid and semi-arid regions are shown in Figure 3.1, using the example of North Africa, where a contrast is drawn between the infrequent and small amounts of modern recharge as compared with the huge reserves of palaeo-waters recharged during wetter climates of the Pleistocene and Holocene. Broad wadi systems and palaeo-lakes testify to former less-arid climates.

Against this background the main water quality issues of arid and semi-arid regions can be considered both for North Africa and other similar regions. Geochemical techniques can be applied for defining and mitigating several of the key water issues in arid and semi-arid areas. These issues include salinisation, recharge assessment, residence time estimation and the definition of natural (baseline) water quality, as a basis for studying pollution and

Groundwater Modelling in Arid and Semi-Arid Areas, ed. Howard S. Wheater, Simon A. Mathias and Xin Li. Published by Cambridge University Press.

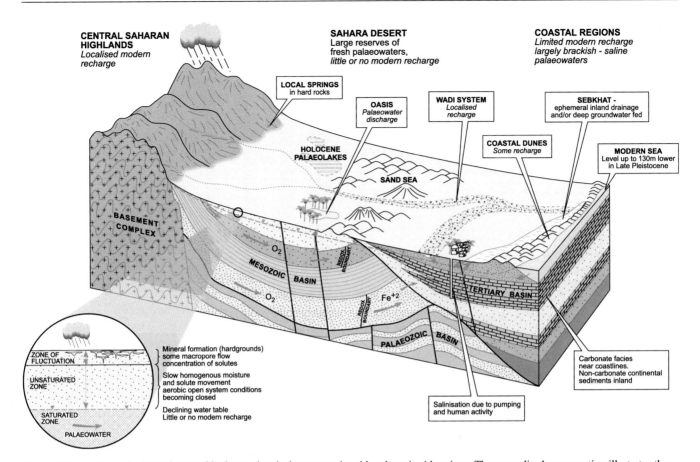

Figure 3.1. Landscape hydrogeology and hydrogeochemical processes in arid and semi-arid regions. The generalised cross-section illustrates the geological environment found in North Africa, where sedimentary basins of different ages overstep each other unconformably and may be in hydraulic continuity with each other. This may be regarded as applicable to many of the areas of the world with large sedimentary basins.

human-induced changes more generally. In this context it is noted that the mainly continental sandstones indicated in Figure 3.1 are highly oxidised and that dissolved O_2 can persist to considerable depths in many basins. Under oxidising conditions several elements (Mo, Cr, As) or species such as NO_3 remain stable and may persist or build up along flow pathways.

This chapter describes the application of hydrogeochemical techniques in conceptualising groundwater quality in semi-arid regions, especially in Africa, and follows the chemical pathway of water from rainfall through the hydrological cycle. In this way it becomes possible to focus on the question of modern recharge, how much is occurring and how to recognise it. It is vital to be able to recognise the interface between modern water and palaeowaters as a basis for sustainable management of water resources in such regions. Chemical and especially isotopic methods can help to distinguish groundwater of different generations and to understand water quality evolution.

3.2 METHODS OF INVESTIGATION

Hydrogeochemical investigations consist of two distinct steps: sampling and analysis. Special care is needed with sampling because of the need for representativeness of sample, the question of mixtures of water (especially groundwaters which may be stratified), the need to filter or not to filter, as well as the stability of chemical species. For groundwater investigations a large range of tools is available, both chemical and isotopic, for investigating chemical processes and overall water quality in arid and semi-arid regions. However, it is essential that unstable variables such as pH, temperature, Eh (redox potential) and DO (dissolved oxygen) be measured in the field. Some details on field approaches are given by Appelo and Postma (1993) and by Clark and Fritz (1997).

A summary of potential techniques for the hydrogeochemical study of groundwaters in arid and semi-arid areas is given in Table 3.1. Chloride can be regarded as a master variable. It is chemically inert and is therefore conserved in the groundwater system – in contrast to the water molecule, which is lost or fractionated during the physical processes of evapotranspiration. The combined use of chloride and the stable isotopes of water ($\delta^{18}O$, δ^2H) therefore provide a powerful technique for studying the evolution of groundwater salinity as well as recharge/discharge relationships (Fontes, 1980; Clark and Fritz, 1997; Coplen *et al.*, 1999). A major challenge in arid-zone investigations is to be able

Table 3.1. *Principal geochemical tools for studies of water quality in arid and semi-arid zones. Examples and references to the literature are given in the text.*

Geochemical/ isotopic tool	Role in evaluating water quality and salinity
Cl	Master variable: inert tracer in nearly all geochemical processes; use in recharge estimation and to provide a record of recharge history.
Br/Cl	Use to determine geochemical source of Cl.
^{36}Cl	Half-life 3.01×10^5 years. Thermonuclear production – use as tracer of Cl cycling in shallow groundwater and recharge estimation. Potential value for dating over long time-spans and also for study of long-term recharge processes. However, in-situ production must be known.
$^{37}Cl/^{35}Cl$	Fractionation in some parts of the hydrological cycle, mainly in saline/hypersaline environments; may allow fingerprinting.
Mg/Ca	Diagnostic ratio for (modern) seawater.
Sr, I, etc.	Diagenetic reactions release incompatible trace elements and may provide diagnostic indicators of palaeomarine and other palaeowaters.
Mo, Cr, As, Fe2+U, NO_3	Metals indicative of oxidising groundwater. Nitrate stable under aerobic conditions.
Fe^{2+}	Indicative of reducing environments.
Nutrients (NO_3, K, PO_4 – also DOC)	Nutrient elements characteristic of irrigation return flows and pollution.
$\delta^{18}O$, δ^2H	Essential indicators, with Cl, of evaporative enrichment and to quantify evaporation rates in shallow groundwater environments. Diagnostic indicators of marine and palaeomarine waters.
^{87}Sr, ^{11}B	Secondary indicators of groundwater salinity source, especially in carbonate environments.
$\delta^{34}S$	Indicator for evolution of seawater sulphate undergoing diagenesis. Characterisation of evaporite and other SO_4. Recognition of modern recharge sources of saline waters.
CFCs, SF_6	Recognition of modern recharge: past 50 years. Mainly unreactive. CFC-12, CFC-11 and CFC-113 used conjunctively.
^3H	Recognition of modern recharge (half-life 12.3 years).
^{14}C, $\delta^{13}C$	Main tool for dating groundwater. Half-life 5730 years. An understanding of carbon geochemistry (including use of ^{13}C) is essential to interpretation.

to distinguish saline water of different origins, including saline build-up from rainfall sources, formation waters of different origins, as well as relict seawater. The Br/Cl ratio is an important tool for narrowing down different sources of salinity (Edmunds, 1996a; Davis *et al.*, 1998), discriminating specifically between evaporite, atmospheric and marine Cl sources. The relative concentrations of reactive tracers, notably the major inorganic ions, must be well understood, as they provide clues to the water–rock interactions which give rise to overall groundwater mineralisation. Trace elements also provide an opportunity to fingerprint water masses; several key elements such as Li and Sr are useful tools for residence time determinations. Some elements such as Cr, U, Mo and Fe are indicative of the oxidation status of groundwater. The isotopes of Cl may also be used: ^{36}Cl to determine the infiltration extent of saline water of modern origin (Phillips, 1999) and $\delta^{37}Cl$ to determine the origins of chlorine in saline formation waters. The measurement precision (better than $\pm0.09‰$) makes the use of chlorine isotopes a potential new tool for studies of environmental salinity (Kaufmann *et al.*, 1993). In addition,

several other isotope ratios, such as $\delta^{15}N$ (Heaton, 1984), $\delta^{87}Sr$ (Yechieli *et al.*, 1992) and $\delta^{11}B$ (Bassett, 1990; Vengosh and Spivak, 1999) may be used to help determine the origins and evolution of salinity. Accumulations of ^4He may also be closely related to crustal salinity distributions (Lehmann *et al.*, 1995).

For most investigations it is likely that conjunctive measurement by a range of methods is desirable (e.g. chemical and isotopic; inert and reactive tracers). However, several of the above tools are only available to specialist laboratories and cannot be widely applied, although it is often found that the results of research using detailed and multiple tools can often be interpreted and applied so that simple measurements then become attractive. In many countries advanced techniques are not accessible and so it is important to stress that basic chemical approaches can be adopted quite successfully. Thus major ion analysis, if carried out with a high degree of accuracy and precision, can prove highly effective; the use of Cl mass balance and major element ratios (especially where normalised to Cl) are powerful investigative techniques.

3.3   TIME-SCALES AND PALAEOHYDROLOGY

Isotopic and chemical techniques are diagnostic for time-scales of water movement and recharge, as well as in the reconstruction of past climates when recharge occurred. In this way, the application of hydrogeochemistry can help solve essentially physical problems and can assist in validating numerical models of groundwater movement.

A palaeohydrological record for Africa has been built up through a large number of palaeolimnological and other archives which demonstrate episodic wetter and drier interludes throughout the late Pleistocene and Holocene (Servant and Servant-Vildary, 1980; Gasse, 2000). The late Pleistocene was generally cool and wet, although at the time of the last glacial maximum much of North Africa was arid, related to the much lower sea surface temperatures. The records show, however, that the Holocene was characterised by a series of abrupt and dramatic hydrological events, mainly related to monsoon activity, although these events were not always synchronous over the continent. Evidence (isotopic, chemical and dissolved gas signatures) contained in dated groundwater forms important and direct proof of the actual occurrence of the wet periods inferred from the stratigraphic record as well as the possible source, temperature and mode of recharge of the groundwater. Numerous studies carried out in North Africa contain pieces of a complex hydrological history.

There is widespread evidence for long-term continuous recharge of the sedimentary basins of North Africa based on sequential changes in radiocarbon activities (Gonfiantini et al., 1974; Edmunds and Wright, 1979; Sonntag et al., 1980). These groundwaters are also distinguishable by their stable isotopic composition; most waters lie on or close to the global meteoric water line (GMWL), but with lighter compositions than at the present day. This signifies that evaporative enrichment was relatively unimportant and that recharge temperatures were lower than at present. This is also supported by noble gas data, which show recharge temperatures at this time typically some 5–7 °C lower than at present (Fontes et al., 1991, 1993; Edmunds et al., 1999). Following the arid period at the end of the Pleistocene, groundwater evidence records episodes of significant recharge coincident with the formation of lakes and large rivers, with significant recharge from around 11 kyr BP. Pluvial episodes with a duration of up to 2000 years are recorded up until around 5.5 kyr BP, followed by onset of the aridity of the present day from around 4.5 kyr BP.

In Libya, during exploration studies in the mainly unconfined aquifer of the Surt basin, a distinct body of very fresh groundwater with concentrations as low as 50 mg l^{-1} (within the zone shown as <500 mg l^{-1} in Figure 3.2) was found to a depth of 100 m which cross-cuts the general NW–SE trend of salinity increase in water dated to the late Pleistocene (Edmunds and Wright, 1979).

This feature (Figure 3.2) is around 10 km in width and may be traced in a roughly NE–SW direction for some 130 km. The depth to the water table is currently around 30–50 m. Because of the good coverage of water-supply wells for hydrocarbon exploration, a three-dimensional impression could be gained of the water quality. It is clear that this feature is a channel that must have been formed by recharge from a former ancient river system. No obvious traces of this river channel were found in the area, which had undergone significant erosion, although Neolithic artefacts and other remains testify that the region had been settled in the Holocene. Whereas the regional more-mineralised groundwater gave values of 0.7–54% modern carbon, indicating a late Pleistocene age, the fresh water gave values of 37.6–51.2% modern carbon (corresponding ages ranging from 5000 to 7800 years) and were also distinctive in their hydrogeochemistry. Evidence from shallow wells in the vicinity, however, proved that fresh water of probable Holocene age was also present more widely at the water table, indicating that direct recharge was simultaneously occurring at a regional scale. Similar channels, preserving the Holocene/Pleistocene interface, have been described by Pearson and Swarzenski (1974) from Kenya. In Mali, the records of Holocene floods from northward migration of the Niger river inland delta are also well recorded (Fontes et al., 1991).

In comparison with the distinctive light isotopic composition of late Pleistocene recharge which lies close to the GMWL, Holocene groundwater is characterised by heavier, more-evaporated compositions (Edmunds and Wright, 1979; Darling et al., 1987; Dodo and Zuppi, 1999). Extrapolation of these light isotopic compositions along lines of evaporation on the delta diagrams to the GMWL allow identification of the composition of the parent rains. It is proposed (Fontes et al., 1993) that these systematically light isotopic compositions were caused by an intensification of the monsoon coincident with northward movement of the inter-tropical convergence zone (ITCZ), resulting in convective heavy rains with low condensation temperatures. From the distribution of such groundwaters over the Sahara it is implied that the ITCZ moved some 500 km to the north during the Holocene.

3.4   RAINFALL CHEMISTRY

Information on the conservative chemical components of rainfall is required by hydrologists to perform chemical mass balance studies of river flow and recharge. Rainfall may also be considered as the 'titrant' in hydrogeochemical processes, since it represents the initial solvent in the study of water–rock interaction. A knowledge of rainfall chemistry can also contribute to our fundamental understanding of air mass circulation, both from the present-day synoptic viewpoint as well as for past climates. In fact, very little information is available on rainfall composition. This is particularly true for Africa, where only limited rainfall

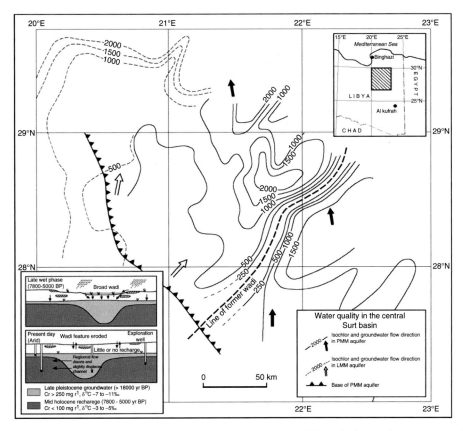

Figure 3.2. Chloride concentration contours in groundwater from the Surt basin, Libya, define a freshwater feature cross-cutting the general water quality pattern. This feature is younger water from a former wadi channel, superimposed on the main body of palaeowater.

chemical data are available. However, a reasonable understanding of the isotopic evolution of rainfall in arid and semi-arid areas is emerging, for example within the African monsoon using isotopes (Taupin *et al.*, 1997, 2000). The stable isotope ($\delta^{18}O$ and δ^2H) composition of precipitation can aid identification of the origin of the precipitated water vapour and its condensation history, and hence give an indication of the sources of air masses and the atmospheric circulation (Rozanski *et al.*, 1993). Several chemical elements in rainfall (notably Cl) behave inertly on entering the soil and unsaturated zone, and may be used as tracers (Herczeg and Edmunds, 1999). The combined geochemical signal may then provide a useful initial tracer for hydrologists, as well as providing a signal of past climates from information stored in the saturated and unsaturated zones (Cook *et al.*, 1992; Tyler *et al.*, 1996; Edmunds and Tyler, 2002). In fact, over much of the Sahara-Sahel region, rainfall-derived solutes, following concentration by evapotranspiration, form a significant component of groundwater mineralisation (Andrews *et al.*, 1994).

Rainfall chemistry can vary considerably in both time and space, especially in relation to distance from the ocean. This is well demonstrated in temperate latitudes for North America (Junge and Werby 1958); the basic relationship between decreasing salinity and inland distance has also been demonstrated

for Australia (Hingston and Gailitis, 1976). The chemical data for Africa are very limited and it is not yet possible to generalise on rainfall chemistry across the continent. The primary source of solutes is marine aerosols dissolved in precipitation, but compositions may be strongly modified by inputs from terrestrial dry deposition. Because precipitation originates in the ocean, its chemical composition near the coast is similar to that of the ocean. Aerosol solutes are dissolved in the atmospheric moisture through release of marine aerosols near the sea surface (Winchester and Duce, 1967). This initial concentration is distinctive in retaining most of the chemical signature of seawater, for example the high Mg/Ca ratio as well as a distinctive ratio of Na/Cl. As rainfall moves inland towards the interior of continents, sulphate and other ions may increase relative to the Na and Cl ions.

The more-detailed studies of African rainfall have used stable isotopes in relation to the passage of the monsoon from its origins in the Gulf of Guinea (Taupin *et al.*, 2000) towards the Sahel, where the air masses track east to west in a zone related to the position and intensity of the inter-tropical convergence zone (ITCZ). While the inter-annual variations and weighted mean compositions of rainfall are rather similar, consistent variations are found at the monthly scale which are related to temperature and relative humidity. This is found from detailed studies in southern Niger, where the isotopic compositions reflect a mixture of recycled vapour (Taupin *et al.*, 1997).

Chloride may be regarded as an inert ion in rainfall, with distribution and circulation in the hydrological cycle taking place solely through physical processes. A study of rainfall chemistry in northern Nigeria has been made by Goni *et al.* (2001). Data collected on an event basis throughout the rainy seasons from 1992 to 1997 showed weighted mean Cl values ranging from 0.6 to 3.4 mg l^{-1}. In 1992, Cl was measured at five nearby stations in Niger with amounts ranging from 0.3 to 1.4 mg l^{-1}. The spatial variability at these locations is notable. However, it seems that, apart from occasional localised convective events, the accumulation pattern for Cl is temporally and spatially uniform during the monsoon; chloride accumulation is generally proportional to rainfall amount over wide areas. This is an important conclusion for the use of chloride in mass balance studies.

Bromide, like chloride, also remains relatively inert in atmospheric processes, though there is some evidence of both physical and chemical fractionation in the atmosphere relative to Cl (Winchester and Duce, 1967). The Br/Cl ratio may thus be used as a possible tracer for air mass circulation, especially to help define the origin of the Cl. The ratio Br/Cl is also a useful palaeoclimatic indicator (Edmunds *et al.*, 1992), since rainfall ratio values may be preserved in groundwater in continental areas. Initial Br enrichment occurs near the sea surface with further modifications taking place over land. The relative amounts of chloride and bromide in atmospheric deposition result initially from physical processes that entrain atmospheric aerosols and control their size. A significant enrichment of Br over Cl, up to an order of magnitude, is found in the Sahelian rains when compared to the marine ratio (Goni *et al.*, 2001). Some Br/Cl enrichment results from preferential concentration of Br in smaller sized aerosol particles (Winchester and Duce, 1967. However, the significant enrichment in the Nigerian ratio is mainly attributed to incorporation of aerosols from the biomass as the monsoon rains move northwards, either from biodegradation or from forest fires. What is clear is that atmospheric dust derived from halite is unimportant in this region, since this would give a much lower Br/Cl ratio.

3.5 THE UNSATURATED ZONE

Diffuse recharge through the unsaturated zone over time-scales ranging from decades to millennia is an important process in controlling the chemical composition of groundwater in arid and semi-arid regions. In this zone fluctuations in temperature, humidity and CO_2 create a highly reactive environment. Below a certain depth (often termed the zero-flux plane) the chemical composition will stabilise, and in homogeneous porous sediments near steady-state movement (piston flow) takes place towards the water table. It is important that measurements of diffuse groundwater recharge only consider data *below* the zero-flux plane. Some vapour transport may still be detectable below these depths at low moisture fluxes, however, as shown by the presence of tritium at the water table in some studies (Beekman *et al.*, 1997). The likelihood of preferential recharge via surface runoff (see below) is also important in many arid and semi-arid regions.

In indurated or heterogeneous sediments in arid and semi-arid systems in some terrains, bypass (macropore or preferential) flow may also be an important process. In older sedimentary formations joints and fractures are naturally present. In some otherwise sandy terrain where carbonate material is present, wetting and drying episodes may lead to mineralisation in and beneath the soil zone, as mineral saturation (especially calcite) is repeatedly exceeded. This is strictly a feature of the zone of fluctuation above the zero-flux plane, however, where calcretes and other near-surface deposits may give rise to hard-grounds with dual porosities. Below a certain depth the pathways of soil macropore movement commonly converge and a more-or-less homogeneous percolation is re-established. In some areas, such as the southern USA, bypass flow via macropores is found to be significant (Wood and Sandford, 1995; Wood, 1999). In areas of Botswana it is found that preferential flow may account for at least 50% of fluxes through the unsaturated zone (Beekman *et al.*, 1999; De Vries *et al.*, 2000).

Four main processes influence soil water composition within the upper unsaturated zone:

1. The input rainfall chemistry is modified by evaporation or evapotranspiration. Several elements, such as chloride (also to a large extent Br, F and NO_3), remain conservative during passage through this zone and the atmospheric signal is retained. A build-up of salinity takes place, although saline accumulations may be displaced annually or inter-annually to the groundwater system.
2. The isotopic signal (δ^2H, $\delta^{18}O$) is modified by evaporation, with loss of the lighter isotope and enrichment in heavy isotopes with a slope of between 3 to 5 relative to the meteoric line (with a slope of 8). Transpiration by itself, however, will not lead to fractionation.
3. Before passage over land, rainfall may be weakly acidic and neutralisation by water–rock interaction will lead rapidly to an increase in solute concentrations.
4. Biogeochemical reactions are important for the production of CO_2, thus assisting mineral breakdown. Nitrogen transformations are also important (see below), leading frequently to a net increase in nitrate input to the groundwater.

These processes may be considered further in terms of the conservative solutes that may be used to determine recharge rates and recharge history. The reactive solutes provide evidence of the controls during reaction, tracers of water origin and pathways of movement, as well as an understanding of the potability of water supplies.

3.5.1 Tritium and ^{36}Cl

Tritium has been widely used in the late twentieth century to advance our knowledge of hydrological processes, especially in temperate regions (Zimmermann *et al.*, 1967). It has also been used in a few key studies in arid and semi-arid zones where it has been applied, in particular, to the study of natural movement of water through unsaturated zones. In several parts of the world including the Middle East (Edmunds and Walton, 1980; Edmunds *et al.*, 1988), North Africa (Aranyossy and Gaye, 1992; Gaye and Edmunds, 1996) and Australia (Allison and Hughes, 1978), classical profiles from the unsaturated zone show well-defined 1960s tritium peaks some metres below surface, indicating homogeneous movement (piston flow) of water through profiles at relatively low moisture contents (2–4 wt%). These demonstrate that low but continuous rates of recharge occur in many porous sediments. In some areas dominated by indurated surface layers, deep vegetation or very low rates of recharge, the tritium peak is less well defined (Phillips, 1994), indicating some moisture recycling to greater depths (up to 10 m), although overall penetration of modern water can still be estimated. Some problems have been created with the application of tritium (and other tracers) to estimate recharge, through sampling above the zero-flux plane, where recycling by vegetation or temperature gradients may occur (Allison *et al.*, 1994).

The usefulness of tritium as a tracer has now largely expired due to weakness of the signal following cessation of atmospheric thermonuclear testing and radioactive decay (half-life 12.3 years). It may still be possible to find the peak in unsaturated zones, but this is likely to be at depths of 10–30 m based on those areas where it has been successfully applied. Other radioisotope tracers, especially ^{36}Cl (half-life 301 000 years), which also was produced during weapons testing, still offer ways of investigating unsaturated zone processes and recharge at a non-routine level. However, in studies where both ^{3}H and ^{36}Cl have been applied, there is sometimes a discrepancy between recharge indications from the two tracers due to the non-conservative behaviour of tritium (Cook *et al.*, 1994; Phillips, 1999). Nevertheless, the position and shape of the tritium peak in unsaturated zone moisture profiles provides convincing evidence of the extent to which 'piston displacement' occurs during recharge, as well as providing reliable estimates of the recharge rate.

An example of tritium profiles (with accompanying Cl profiles) sampled in 1977 from adjacent sites in Cyprus is shown in Figure 3.3, which shows one well-defined peak at 10–12 m depth in AK3 and the lower limb of a peak also in AK2. These peaks correspond with rainfall tritium (uncorrected) for the period from

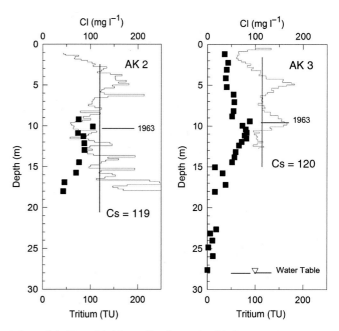

Figure 3.3. Two chloride profiles from Akrotiri, Cyprus, with corresponding tritium (solid squares). Cs is mean (steady-state) chloride in the profile in mg l^{-1}.

the 1950s to the 1970s. The profiles approximate piston displacement and the peaks align well. A vertical moisture movement of about 0.7 m yr^{-1} is calculated and the recharge rates are 52 and 53 mm yr^{-1}; these figures compare well with estimates using Cl (Edmunds *et al.*, 1988). A further example of tritium as a time-scale indicator is shown in Figure 3.4.

3.5.2 Stable isotopes

Stable isotopes have been used in the study of recharge but in general only semi-quantitative recharge estimates can be obtained. In arid and semi-arid regions, however, the fractionation of isotopes between water and vapour in the upper sections of unsaturated zone profiles has been used successfully to calculate rates of discharge (Barnes and Allison, 1983) and to indicate negligible recharge (Zouari *et al.*, 1985).

At high rainfall, the recharging groundwater undergoes seasonal fractionation within the zone of fluctuation (Darling and Bath, 1988), but any seasonal signal is generally smoothed out with the next pulse of recharge and little variation remains below the top few metres. In arid and semi-arid zones, however, where low recharge rates occur, the record of a sequence of drier years may be recorded as a pulse of ^{18}O-enriched water, as recorded for example from Senegal (Gaye and Edmunds, 1996). Isotopic depletion in deep unsaturated zones in North America has been used to infer extended wet and dry intervals during the late Pleistocene (Tyler *et al.*, 1996). Extreme isotopic enrichment in the unsaturated zone accompanies chloride accumulation over intervals when recharge rates are zero (Darling *et al.*, 1987).

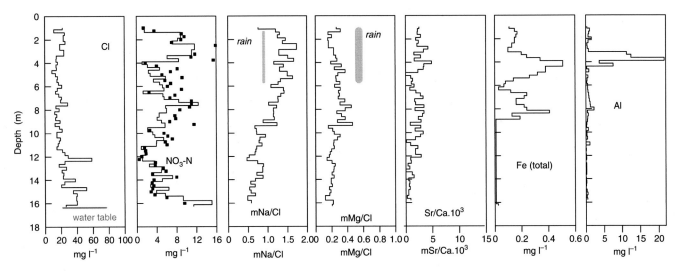

Figure 3.4. Geochemistry of interstitial waters in the unsaturated zone – profile MD1 from northern Nigeria with concentrations and ratios for a number of elements. The molar N/Cl ($\times 10$) is also plotted on the nitrate diagram.

3.5.3 Chloride

Numerous studies using Cl as a conservative tracer in recharge calculations have been reported, and Cl mass-balance methods probably offer the most reliable approach to recharge estimation for semi-arid and arid regions (Allison *et al.*, 1994). In addition, Cl analysis is inexpensive and is widely applicable, bringing it within the budgets of most water resource organisations, although the capacity for accurate measurements at low Cl concentrations is required. The methods of investigation are straightforward and involve the recovery of dry samples by augur, percussion drilling or dug wells, followed by the measurement of moisture content and the elution of Cl. A number of criteria must be satisfied or taken into account for successful application of the technique: that no surface runoff occurs; that Cl is solely derived from rainfall; that Cl is conservative with no additions from within the aquifer; and that steady-state conditions operate across the unsaturated interval where the method is applied (Edmunds *et al.*, 1988; Herczeg and Edmunds, 1999; Wood, 1999). As with tritium, it is important that sampling is made over a depth interval which passes through the zone of fluctuation. For the example shown in Figure 3.3, from Cyprus, the mean (steady-state) concentrations of Cl in the pore waters (119 and 122 mg l^{-1} respectively) provide recharge estimates of 56 and 55 mm yr^{-1}. These estimates are derived using mean Cl data over three years at the site for rainfall chemistry and the measured moisture contents. Macropore flow is often present in the soil zone, but in the Cyprus and northern African profiles in the absence of large trees this is restricted to the top 2–4 m of the profile.

3.5.4 Nitrate

Groundwater beneath the arid and semi-arid areas of the Sahara is almost exclusively oxidising. The continental aquifer systems contain little or no organic matter, and dissolved oxygen concentrations of several mg l^{-1} may persist in palaeowater dated in excess of 20 000 years old (Winograd and Robertson, 1982, Heaton *et al.*, 1983). Under these conditions nitrate also is stable and acts as an inert tracer recording environmental conditions. Nitrate concentrations in Africa are often significantly enriched and frequently exceed 10 mg l^{-1} NO$_3$-N (Edmunds and Gaye, 1997; Edmunds, 1999). High nitrate concentrations may be traced in interstitial waters through the unsaturated zone to the water table and are clearly the result of biogeochemical enrichment. This enrichment is well above concentrations from the atmosphere, allowing for evapotranspiration, and the source is most likely to be naturally occurring N-fixing plants such as acacia, though in the modern era some cultivated species may also contribute. In Senegal (Edmunds and Gaye, 1997), the NO$_3$-N/Cl ratio can even exceed 1.0. A record of nitrate in the unsaturated zone can therefore be used as an indicator of vegetation changes, including land clearance and agriculture.

3.5.5 Reactive tracers and water–rock reactions in the unsaturated zone

The main process of groundwater mineralisation takes place by acid–base reactions in the top few metres of the Earth's crust. Natural rainfall acidity is between pH 5 and 5.5. Atmospheric CO$_2$ concentrations are low, but may increase significantly due to microbiological activity in the soil zone. In warmer tropical latitudes the solubility of CO$_2$ is lower than in cooler mid-high latitude areas and weathering rates will be somewhat lower (Tardy, 1970; Berner and Berner, 1996). The concentration of soil CO$_2$ (pCO$_2$) is nevertheless of fundamental importance in determining the extent of a reaction; an open system with respect to the soil/atmosphere reservoir may be maintained by diffusion to a depth of several metres.

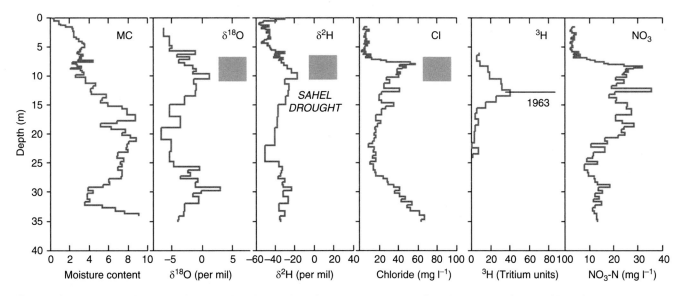

Figure 3.5. Profiles of tritium, stable isotopes, chloride and nitrate in the unsaturated zone from the same location – profile L18, Louga, Senegal.

Investigation of the geochemistry of pore solutions in the unsaturated zone is usually hampered by low moisture contents. Elutriation using distilled water is possible for inert components but not for reactive solutes, since artefacts may be created in the process. Where moisture contents exceed about 5 wt% it may be possible to extract small volumes by immiscible liquid displacement (Kinniburgh and Miles, 1983).

3.5.6 Recharge history

In deep unsaturated zones where piston flow can be demonstrated and where recharge rates are very low it may be possible to reconstruct climate and recharge history. Such profiles have been used to examine the recharge history over the decadal to century scale (or longer) in several parts of the world, as well as to reconstruct the related climate change (Edmunds and Tyler, 2002). More recently global estimates of recharge have been made based mainly on Cl mass balance and these are examined in relation to climate variability (Scanlon et al., 2006). Records up to several hundreds of years have been recorded from China (Ma and Edmunds, 2005) and for several millennia from USA (Tyler et al., 1996).

An example of an integrated study from Senegal using tritium, Cl and stable isotopes is given in Figure 3.5, based on Gaye and Edmunds (1996) and Aranyossy and Gaye (1992). Samples were obtained from Quaternary dune sands where the water table was at 35 m and where the long-term (100 year) average rainfall is 356 mm yr^{-1} (falling by 36% to 223 mm since 1969 during the Sahel drought). The tritium peak clearly defines the 1963 rainfall as well as demonstrating that piston displacement is occurring. A recharge rate of 26 mm yr^{-1} is indicated and little if any bypass flow is taking place.

The profile has a mean Cl concentration (23.6 mg l^{-1}) that corresponds to a recharge rate of 34.4 mm yr^{-1}, calculated using a value for rainfall Cl of 2.8 mg l^{-1} and an average rainfall of 290 mm yr^{-1}. This rate is in close agreement with the tritium-derived value, indicating homogeneous movement with little dispersion of both water and solute. Oscillations in Cl with depth are due to variable climatic conditions, and the prolonged Sahel drought is indicated by the higher Cl concentrations. These profiles have been used to create a chronology of recharge events (Edmunds et al., 1992). The stable isotope values are consistent with the trends in recharge rates indicated by Cl, the most enriched values corresponding to high Cl and periods of drought, but the stable isotopes are unable to quantify the rate of recharge, unlike Cl and ^3H.

3.5.7 Spatial variability

The use of the chloride mass balance in the unsaturated zone is now being widely adopted for *diffuse recharge* estimation in semi-arid regions (Allison et al., 1994; Gaye and Edmunds, 1996; Wood and Sandford, 1995). Conservative rainfall solutes, especially chloride, are concentrated by evapotranspiration and the consequent build-up in salinity is inversely proportion to the recharge rate. This method is therefore ideally suited to the measurement of low recharge rates; it is also attractive since, in sharp contrast to instrumental approaches, decadal or longer average rates of recharge, preserved in thick unsaturated zone profiles, may be obtained. The results of recent applications of the chloride mass balance approach using profiles to derive recharge assessments in different parts of the northern hemisphere (Africa and Asia) are shown in Figure 3.6. Some of these are single data points but for some areas multiple sites are

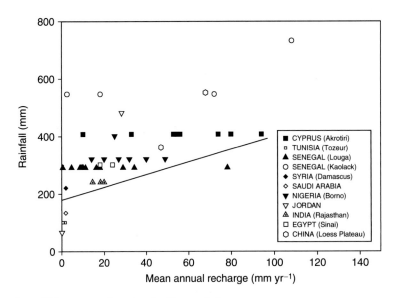

Figure 3.6. Recharge estimates from different countries using the Cl mass balance.

represented for the same area; only those profiles are shown which are ≥ 8 m in length, where steady-state moisture and chloride are recorded, indicating net downward flux of moisture. The sources of data are given in the diagram. Several interesting features emerge:

1. There is a cutoff point around 200 mm mean annual rainfall which signifies the lower limit for effective recharge.
2. Above this threshold there is the possibility for recharge as in Cyprus where 100 mm recharge is measured from 400 mm rainfall. Nevertheless, the technique records small but finite recharge rates in some areas with rainfall as high as 550 mm.
3. At any one site a large range of recharge may occur in a small area implying that spatial variability is likely to be considerable, due to soil, geology or vegetation controls, since seasonal rainfall variability is likely to be evened out over the decadal time-scales of most profiles.
4. Finite, if low, recharge rates still occur even in those areas with mean annual rainfall below 200 mm (e.g. Tunisia, Saudi Arabia). However, these low rates also signify moderate salinity in the profiles; with the absence of recharge, salinity accumulation will be continuous as found in southern Jordan.

Most of the profiles are from sandy terrain (sandstones or unconsolidated dune sands). Few examples are shown of clay-rich sediments. An exception is the Damascus basin profile which is predominantly clay-loam and which shows very low recharge. Two profiles from China are from loess sediments and these show modest rates of recharge. The recharge phenomenon is examined further with reference to the data from Senegal where a wide range of recharge rates has been measured across the Sahel region with rainfall ranging from 290–720 mm yr^{-1}.

One way to determine the spatial variability of recharge using geochemical techniques is to use data from shallow dug wells. Water at the immediate water table in aquifers composed of relatively homogeneous porous sediments represents that which is derived directly from the overlying unsaturated zone. Recharge estimates may then be compared with those from unsaturated-zone profiles to provide robust long-term estimates of renewable groundwater, taking into account the three dimensional moisture distributions and based on time-scales of rainfall from preceding decades. A series of 120 shallow wells in a 1600 km^2 area of northwest Senegal (Figure 3.7) were used to calculate the areal recharge from the same area from which the point source recharge estimates were obtained (Edmunds and Gaye, 1994). Average chloride concentration for seven separate profiles at this site is 82 mg l^{-1}, giving a spatially averaged recharge of 13 mm yr^{-1}.

The topography of this area has a maximum elevation of 45 m and decreases gently towards the coast and to the north; the natural groundwater flow is to the northwest. The dug wells are generally outside the villages and human impacts on the water quality are regarded as minimal; the area is either uncultivated savannah or has been cleared and subjected to rainfed agriculture. The Cl concentrations in the shallow wells have been contoured, and range from < 50 to > 1000 mg l^{-1} Cl (Figure 3.7a). These are proportional to recharge amounts which vary from > 20 to < 1 mm yr^{-1} (Figure 3.7b) The spatial variability may be explained mainly by changes in sediment grain size between the central part of the dune field and its margins where the sediments thin over older less-permeable strata. Over the whole area the renewable resource may be calculated at between 1100 and 13 000 m^3 km^{-2} yr^{-1}.

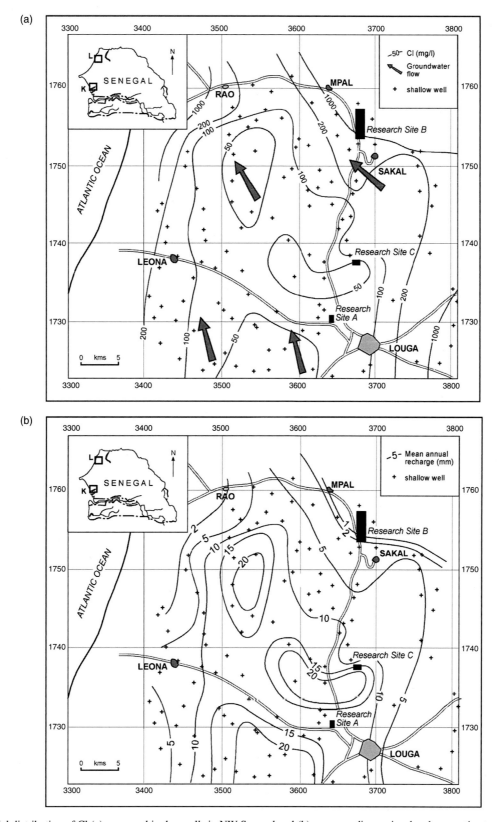

Figure 3.7. Spatial distribution of Cl (a) measured in dug wells in NW Senegal and (b) corresponding regional recharge estimates. Unsaturated zone profiles were measured at the research sites indicated and gave good correlation for diffuse recharge rates.

3.5.8   Focused recharge

The chapter has focused so far on diffuse recharge where regional resource renewal takes place through mainly homogeneous sandy terrain, typical of much of North Africa. In areas of hard rock or more clay-rich terrain or where there is a well-developed drainage system, it is likely that recharge will be focused along wadi lines and some recharge will occur via storms and flash floods, even where the mean annual average rainfall is below 200 mm.

One example of this has been described from Sudan (Darling et al., 1987; Edmunds et al., 1992). The wadi system (Wadi Hawad) flows as a result of a few heavy storms each year in a region with rainfall around 200 mm yr^{-1}. The regional terrain (the Butana Plain) is mainly clay sediments and it can be shown that there is no net diffuse recharge, demonstrated by the fact that Cl is accumulating in the unsaturated zone. The deep groundwater in the region is clearly a palaeowater with recharge having occurred in the Holocene. Shallow wells at a distance of up to 1 km from the wadi line contain water with tritium, all of which indicated a large component of recent recharge (24–76 TU). Samples of deep wells from north of the area, away from the main Wadi Hawad, all gave values $\leq$ 9 TU. Thus the wadi acts as a linear recharge source and sustains the population of small villages and one small town through a freshwater lens that extends laterally from the wadi.

3.6   HYDROCHEMISTRY OF GROUNDWATER SYSTEMS IN ARID AND SEMI-ARID REGIONS

From the foregoing discussion it is evident from geochemical and isotopic evidence that significant diffuse recharge to aquifer systems does not occur at present in most arid and semi-arid regions. In the Senegal example above, recharge rates to sandy areas with long-term average rainfall of around 350 mm are between <1 and 20 mm. It is likely that these results could be extrapolated to other regions with similar landscape, geology and climate over much of northern Africa, though preferential recharge along drainage systems is also important, providing areas with small but sustainable supplies. These present-day low recharge conditions will have an impact on the overall chemistry due to lower water–rock ratios and larger residence times, leading to somewhat higher salinities and evolved hydrogeochemistry.

Africa is characterised by several large overstepping sedimentary basins (Figure 3.1) which contain water of often excellent quality. Much of this water is shown to be palaeowater recharged during the late Pleistocene or Holocene pluvial periods. Hydrogeochemical processes observed in these aquifers have principally been taking place along flowlines under declining heads towards discharge points in salt lakes or at the coast. Present rates of groundwater movement in the large sedimentary basins are likely to be less than one metre per year; in the Great Artesian Basin of Australia rates as low as 0.25 mm yr^{-1} are recorded (Love et al., 2000). Piezometric decline has generally been occurring, since fully recharged conditions and transient conditions are likely to be still operating at the present day as adjustment continues towards modern recharge conditions. Groundwater in large basins is never 'stagnant' and water–rock interaction will continue to occur. The evolution of inorganic groundwater quality is considered here in the light of the isotopic evidence, which provides information on the age and different recharge episodes in the sedimentary sequences.

3.6.1   Input conditions – inert elements and isotopic tracers

The integrated use of inert elements and chemical and isotopic techniques provides a powerful means of determining the origin (s) of groundwater in semi-arid regions. To a large extent these tracers (Cl, δ^{18}O, δ^2H) are the same as used in studies of the unsaturated zone and may be adopted to clearly fingerprint water from past recharge regimes. However, the longer time-scales of groundwater recharge require the assistance of radiocarbon dating and, if possible, other radioisotopic tools such as ^{36}Cl, ^{81}Kr and ^{39}Ar (Loosli et al., 1999). Noble gas isotopic ratios also provide evidence of recharge temperatures which are different from those of the present day (Stute and Schlosser, 1999).

Several elements, especially Br and Br/Cl ratios, remain effectively inert in the flow system and may be used to follow various input sources and the evolution of salinity in aquifers (Edmunds, 1996a; Davis et al., 1998). An example is given from the Continental Intercalaire in Algeria (Edmunds et al., 2003) along a section from the Saharan Atlas to the discharge area in the Tunisian Chotts (Figure 3.8). The overall evolution is indicated by Cl, which increases from around 200 to 800 mg l^{-1}. The sources of salinity are shown by the Br/Cl ratio; the increase along the flow lines for some 600 m is due to the dissolution of halite from one or more sources, while in the area of the salt lakes (chotts) some influence of marine formation waters is indicated (possibly from water flowing to the chotts from a second flow line). Fluorine remains buffered at around 1 mg l^{-1} controlled by saturation with respect to fluorite and iodine follows fluorine behaviour, being released from the same source as F possibly from organic rich horizons. The Cl and Br concentrations and ratio are related to inputs and differ from F and I, which are principally controlled by water–rock interaction.

3.6.2   Reactions and evolution along flow lines

The degree of groundwater mineralisation in the major water bodies in semi-arid and arid zones, recharged during the

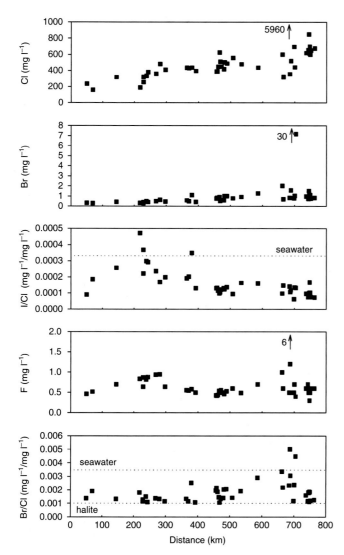

Figure 3.8. Down-gradient hydrogeochemical profile (west-to-east) in the Continental Intercalaire aquifer from the Atlas mountains of Algeria to the discharge area in the Chotts of southern Tunisia. The chloride profile indicates a progressive increase in salinity. The other halogen elements Br, I, F help to explain the origins of salinity and the general chemical evolution down-gradient.

Holocene and the late Pleistocene, is likely to be a reflection of atmospheric/soil chemical inputs, the relatively high recharge rates during the pluvial episodes, the aquifer sedimentary facies and mineralogy. Most of North Africa south of latitude 28° N comprises variable thicknesses of continental sediments overlying crystalline basement, whereas to the north marine facies may occur containing residual formation waters and intra-formational evaporites which lead to high salinities. Similar relationships are found in the coastal aquifers of West Africa, as well as on the Gulf of Guinea. In the Surt basin of Libya the change of facies to marine sediments is marked by a progressive change in chemistry (especially Sr increase) as groundwater moves north of latitude 28° 30′ (Edmunds, 1980). For many of

the large basins of the Sahara and Sahel, away from coastal areas, the groundwater chemistry for inert solutes reflects, to a significant extent, inputs from the atmosphere (Fontes *et al.*, 1993). Superimposed on these inputs, the effects of water–rock interactions in mainly silicate-dominated rock assemblages are recorded.

Important changes in reactive tracers (major and trace elements) also occur, which can be followed by (time-dependent) water–rock reactions. In addition to the chemical signature derived from atmospheric inputs, the chemistry of groundwater is determined to a significant extent by reactions taking place in the first few metres of the unsaturated or saturated zone and reflecting the predominant rock type. Any incoming acidity will be neutralised by carbonate minerals or, if absent, by silicate minerals. In the early stages of flow the groundwater will approach saturation with carbonates (especially calcite and dolomite) and thereafter will react relatively slowly with the matrix in reactions where impurities (e.g. Fe, Mn and Sr) are removed from these minerals, and purer minerals are precipitated under conditions of dynamic equilibrium towards saturation limits with secondary minerals (e.g. fluorite and gypsum).

During this stage, other elements may accumulate with time in the groundwater and indicate if the flow process is homogeneous or not; discontinuities in the chemistry are likely to indicate discontinuities in the aquifer hydraulic connections. An example from the Continental Intercalaire aquifer in Algeria/Tunisia (Edmunds *et al.*, 2003) shows how the major elements may vary along the flow line. The ratios of reactive versus the inert tracer (Cl) are used to indicate the evolution. The very constant Na/Cl ratio (weight ratio of 0.65) throughout the flow path as salinity increases (shown by Cl) indicates the dissolution of halite with very little reaction; at depth a different source is indicated by the higher salinity. The Mg/Ca ratio indicates that saturation with respect to calcite (0.60) is approached after a short distance but that this is disturbed at depth by water depleted in Mg, probably from the dissolution of gypsum. The potassium (K) and K/Cl ratio increase along the flow line from 0.05 to 0.18, suggesting a time-dependent release of K from feldspars or other silicate sources.

3.6.3 Redox reactions

Under natural conditions, groundwater undergoes oxidation-reduction (redox) changes moving along flow lines (Champ *et al.*, 1979; Edmunds *et al.*, 1984). The solubility of oxygen in groundwater at the point of recharge (around 10–12 mg l^{-1}) reflects the ambient air temperature and pressure. The concentration of dissolved oxygen (DO) in newly recharged groundwater may remain high (8–10 mg l^{-1}), indicating relatively little loss of DO during residence in the soil or unsaturated zone (Edmunds *et al.*, 1984). Oxygen slowly reacts with organic matter and/or with Fe^{2+} released from dissolution of impure carbonates, ferromagnesian silicates or with sulphides. Oxygen may persist, however, for many thousands of years in unreactive sediments (Winograd and Robertson, 1982). Complete reaction of oxygen is marked by a fall in the redox potential (Eh) by up to 300 mV. This provides a sensitive index of aquifer redox status. A decrease in the concentration of oxygen in pumped groundwater therefore may herald changes in the input conditions. An increase in dissolved iron (Fe^{2+}) concentration as well as the disappearance of NO_3 may be useful as secondary indicators.

The Continental Intercalaire aquifer illustrates redox conditions for a typical continental aquifer system in an arid zone. The redox boundary corresponds with a position some 300 km along the flow path in the confined aquifer and in waters of late Pleistocene age (Edmunds et al., 2003). Neither Eh nor O_2 measurements were possible in the study, but the position of the redox change is clearly marked by the disappearance of NO_3, the increase in Fe^{2+}, as well as in the sharp reduction in concentrations of several redox-sensitive metals (Cr, U for example), which are stabilised as oxy-anions under oxidising conditions.

3.6.4 Salinity generation

Salinity build-up in groundwater in arid and semi-arid regions has several origins, some of which are referred to in the preceding sections. Edmunds and Droubi (1998) review the topic and discuss the main issues and techniques for studies in such regions. Three main sources are important: atmospheric aerosols, seawater of various generations and evaporite sequences, all of which may be distinguished using a cocktail of chemical or isotopic techniques (Table 3.1).

Atmospheric inputs that slowly accumulate over geological time-scales are of great importance as a source of salinity. The impact on groundwater composition will be proportional to the inputs (including proximity to marine or playa source areas), climate change sequences and the turnover times of the groundwater bodies. The accumulation of salinity from atmospheric sources is most clearly demonstrated from the Australian continent (Hingston and Galitis, 1976), where the landscape and groundwaters reflect closely the deposition of aerosols over past millennia.

Formation waters from marine sediments are important as a salinity source in aquifers near to modern coastlines. Different generations of salinity may be recognised, either from formation waters laid down with young sediments or from marine incursions arising from eustatic or tectonic changes. Evaporites containing halite and/or gypsum are an important cause of quality deterioration in many aquifers in present-day arid and semi-arid regions. Formation evaporites are usually associated with inland basins of marine or non-marine origin. The different origins and generations of salinity in formation waters may be characterised by a range of isotopic and chemical tracers, such as $\delta^{87}Sr$ and $\delta^{11}B$ (Table 3.1).

3.7 CONCLUSIONS

This chapter has focused on water quality from a number of different viewpoints: (a) as a diagnostic tool for understanding the occurrence evolution of groundwater; (b) as a means for determining the recharge amounts; and (c) as a means for studying the controls on groundwater quality dependent on geology and time-scales. Emphasis has been placed on natural processes rather than direct human impacts such as pollution. Nevertheless, in arid regions the deterioration of groundwater quality due to human activity, such as excessive pumping, may lead to loss of resources through salinisation and the approach adopted here will assist in a better understanding of what may be going on. Salinity increase is also a consequence of point source (single well/village) or regional (town and city) pollution, and any overall quality deterioration may be monitored using simple parameters such as specific electrical conductance (SEC) or chloride.

The approach adopted here has been to recognise that water contains numerous items of chemical information that may be interpreted to aid water management. Measurement in the field and in the laboratory of the basic chemistry can inexpensively provide additional information during exploration, development and monitoring of groundwater resources. Too often only physical parameters are seen as important. For example, it is important to understand not only that water levels are falling, but why they are falling. Some key recommendations that emerge from this chapter may be summarised as follows.

3.7.1 Data requirements

Useful information may be obtained through the measurement and careful interpretation of a few key parameters. In the field, the measurement of T, pH and SEC are thus recommended. Temperature is essential, since it provides a proxy for the depth of sample origin in cases where details of production are unknown. Field measurement of pH is essential if meaningful interpretation of the results, especially for groundwater modelling, is to be made, since loss of CO_2 will take place between the well head and the laboratory, leading to a pH rise; laboratory pH is therefore meaningless. SEC is a key parameter in both surveys and monitoring, enabling the hydrogeologist to quickly detect spatial or temporal changes in the groundwater. In the field, two filtered (0.45 μm) samples (one acidified and the other not acidified) can then be taken for laboratory analysis (Edmunds, 1996b).

In hydrochemical studies the measurement of chloride is particularly important, since Cl behaves as an inert tracer which allows it to be used as a reference element to follow reactions in the aquifer and to interpret physical processes such as groundwater recharge, mixing and the development of salinity. Chloride should always be measured, and it is recommended that rainfall Cl also be measured in collaboration with meteorologists.

Most central laboratories are able to cope with at least major ion analyses and these should be measured as the basic data set for water quality investigations, after first establishing that there is an ionic balance (Appelo and Postma, 1993). Minor and trace elements described above are helpful in the diagnosis of processes taking place in the aquifer as well as in determining potability criteria. The more detailed measurements described in this chapter are desirable, but are not essential for routine investigations. Nevertheless, the implications of case studies where combined isotopic and chemical data have been used should be digested and extrapolated for local studies. The results of expensive and specialised case studies are often of considerable value, since at practical and local level these then allow simple tools such as nitrate or chloride to be subsequently used for monitoring.

3.7.2 Groundwater resources assessment

In arid and semi-arid regions the main resources issue is the amount of modern recharge. In this chapter stress has been placed on how to determine whether modern recharge is taking place and, if so, how much. In sandy terrain, or other areas with unconsolidated sediments, it is recommended that attention be paid to information contained in the unsaturated zone. The use of Cl to measure the direct recharge component is recommended. However, the distribution of Cl on a regional scale is also important, since salinity variations will be proportional to the amount of recharge. The use of isotopes and additional chemical parameters will then be needed to help confirm whether this recharge belongs to the modern cycle or is palaeowater.

3.7.3 Groundwater exploration and development

Important information can be obtained during the siting and completion of new wells or boreholes. Chemical information obtained at the end of pumping tests provides the 'initial' baseline chemistry against which subsequent changes may be monitored. It is recommended that analyses of all major ions, minor ions and also stable isotopes of oxygen and hydrogen, if possible, be carried out during this initial commissioning phase.

In more extensive groundwater development it is recommended that exploration of the quality variation with depth be made. One or more boreholes should be regarded as exploration wells for this purpose. Depth information can be obtained during the drilling using packer testing, air-lift tests or similar, which can then help determine the optimum screen design or well depth, for example to minimise salinity problems. Alternatively, single wells can be drilled to different depths in a well field for purposes of studying the stratification in quality or in age (e.g. whether modern groundwater overlies palaeowater). It should be remembered that subsequent pumped samples are inevitably going to be mixtures in terms of age and quality, and this initial testing will enable assessment of changes to be made as the wells are used.

3.7.4 Groundwater quality and use

Most water sampling programmes are heavily oriented towards suitability for drinking use, having regard to health-related problems. In this chapter emphasis has been placed on understanding the overall controls on water quality evolution. Natural geochemical reactions taking place over hundreds or thousands of years will thus give rise to distinctive natural properties. It is important to recognise this natural baseline, without which it will be impossible to identify if pollution from human activity is taking place. It has also been shown how salinity distribution is related to natural geological and climatic factors. In addition, the oxidising conditions prevalent in many arid and semi-arid regions may give rise to enhanced concentrations of metals such as Cr, As, Se and Mo. Prolonged residence times may also lead to high fluoride and manganese concentrations. Under reducing conditions, high Fe concentrations may occur.

REFERENCES

Allison, G. B. and Hughes, M. W. (1978) The use of environmental chloride and tritium to estimate total recharge to an unconfined aquifer. *Aust. J. Soil. Res.* **16**, 181–195.

Allison, G. B., Gee, G. W. and Tyler, S. W. (1994) Vadose-zone techniques for estimating groundwater recharge in arid and semi-arid regions. *Soil Sci. Soc. Am. J.* **58**, 6–14.

Andrews, J. N., Fontes, J. Ch., Edmunds, W. M., Guerre, A. and Travi, Y. (1994) The evolution of alkaline groundwaters in the Continental Intercalaire aquifer of the Irhazer Plain, Niger. *Water Resour. Res.* **30**, 45–61.

Appelo, C. A. J. and Postma, D. (1993) *Geochemistry, Groundwater and Pollution.* Balkema.

Aranyossy, J. F. and Gaye, C. B. (1992) La recherche du pic du tritium thermonucléaire en zone non-saturée profonde sous climat semi-aride pour la mesure de recharge des nappes: première application au Sahel. *C.R. Acad. Sci. Paris*, **315** (ser. II), 637–643.

Barnes, C. J. and Allison, G. B. (1983) The distribution of deuterium and ^{18}O in dry soils. 1. Theory. *J. Hydrol.* **60**, 141–156.

Bassett, R. L. (1990) A critical evaluation of the available measurements for the stable isotopes of boron. *Appl. Geochem.* **5**, 541–554.

Beekman, H. E., Selaolo, E. T. and De Vries, J. J. (1997) *Groundwater Recharge and Resources Assessment in the Botswana Kalahari.* GRES-II project final report, Lobatse.

Beekman, H. E., Selaolo, E. T. and De Vries, J. J. (1999) *Groundwater Recharge and Resources Assessment in the Botswana Kalahari.* Executive Summary GRES-II, Geological Survey of Botswana and Faculty of Earth Sciences, Vrije Universiteit, Amsterdam.

Berner, E. K. and Berner, R. A. (1996) *Global Environment: Water, Air and Geochemical Cycles.* Prentice Hall.

Champ, D. R., Gulens, J. and Jackson, R. E. (1979) Oxidation-reduction sequences in groundwater flow systems. *Canadian J. Earth Sci.* **16**, 12–23.

Clark, I. and Fritz, P. (1997) *Environmental Isotopes in Hydrogeology.* Lewis, Boca Raton.

Cook, P. G., Edmunds, W. M. and Gaye, C. B. (1992) Estimating palaeorecharge and palaeoclimate from unsaturated zone profiles. *Water Resour. Res.* **28**, 2721–2731.

Cook, P. G., Jolly, I. D., Leaney, F. W., Walker, G. R., Allan, G. L., Fifield, L. K. and Allison, G. B. (1994) Unsaturated zone tritium and chlorine-36 profiles from southern Australia: their use as tracers of soil water movement. *Water Resour. Res.* **30**, 1709–1719.

Coplen, T. B., Herczeg, A. L. and Barnes, C. (1999) Isotope engineering – using stable isotopes of the water molecule to solve practical problems. In *Environmental Tracers in Subsurface Hydrology*, ed. P. G. Cook and A. L. Herczeg, 79–110, Kluwer.

Darling, W. G. and Bath, A. H. (1988) A stable isotope study of recharge processes in the English Chalk. *J. Hydrology* **101**, 31–46.

Darling, W. G., Edmunds, W. M., Kinniburgh, D. G. and Kotoub, S. (1987) Sources of recharge to the Basal Nubian Sandstone Aquifer, Butana Region, Sudan. In *Isotope Techniques in Water Resources Development*, 205–224, IAEA.

Davis, S. N., Whittemore, D. O. and Fabryka-Martin, J. (1998) Uses of chloride/bromide ratios in studies of potable water. *Ground Water* **36**, 338–350.

De Vries, J. J., Selaolo, E. T. and Beekman, H. E. (2000) Groundwater recharge in the Kalahari, with reference to paleo-hydrologic conditions. *J. Hydrology* **238**, 110–123.

Dodo, A. and Zuppi, G. M. (1999) Variabilité climatique durant le Quaternaire dans la nappe du Tarat (Arlit, Niger). *C. R Acad. Sci. Paris* **328**, 371–379.

Edmunds, W. M. (1980) The hydrogeochemical characteristics groundwaters in the Sirte Basin, using strontium and other elements. In *Geology of Libya*, Vol. 2, 703–714, University of Tripoli.

Edmunds, W. M. (1996a) Bromide geochemistry in British groundwaters. *Mineral Mag.* **60**, 275–284.

Edmunds, W. M. (1996b) Geochemical framework for water quality studies in sub-Saharan Africa. *J. African Earth Sciences* **22**, 385–389.

Edmunds, W. M. (1999) Groundwater nitrate as a palaeoenvironmental indicator. In *Geochemistry of the Earth's Surface* (Proc. 5th International Symposium on the Geochemistry of the Earth's Surface Reykjavik, Iceland, 35–38. Balkema.

Edmunds, W. M. and Droubi, A. (1998) Groundwater salinity and environmental change. In *Isotope Techniques in the Study of Past and Current Environmental Changes in the Hydrosphere and Atmosphere*, 503–518. IAEA.

Edmunds, W. M. and Gaye, C. B. (1994) Estimating the spatial variability of groundwater recharge in the Sahel using chloride. *J. Hydrol.* **156**, 47–59.

Edmunds, W. M. and Gaye, C. B. (1997) High nitrate baseline concentrations in groundwaters from the Sahel. *J. Environmental Quality* **26**, 1231–1239.

Edmunds, W. M. and Tyler, S. W. (2002) Unsaturated zones as archives of past climates: towards a new proxy for continental regions. *Hydrogeol. J.* **10**, 216–228.

Edmunds, W. M. and Walton, N. R. G. (1980) A geochemical and isotopic approach to recharge evaluation in semi-arid zones, past and present. In *Arid Zone Hydrology, Investigations with Isotope Techniques*, 47–68. IAEA.

Edmunds, W. M. and Wright, E. P. (1979) Groundwater recharge and palaeoclimate in the Sirte and Kufra basins, Libya. *J. Hydrol.* **40**, 215–241.

Edmunds, W. M, Miles, D. L. and Cook, J. M. (1984) A comparative study of sequential redox processes in three British aquifers. In *Hydrochemical Balances of Freshwater Systems* (Proc. Uppsala Symposium, Sept. 1984), ed. E. Eriksson, IAHS Publ. **150**, 55–70.

Edmunds, W. M., Darling, W. G. and Kinniburgh, D. G. (1988) Solute profile techniques for recharge estimation in semi-arid and arid terrain. In *Estimation of Natural Groundwater Recharge*, 139–158. NATO ASI Series, Reidel, Dordrecht.

Edmunds, W. M., Gaye, C. B. and Fontes, J -Ch. (1992) A record of climatic and environmental change contained in interstitial waters from the unsaturated zone of northern Senegal. In *Isotope Techniques in Water Resources Development*, 533–549. IAEA.

Edmunds, W. M., Fellman, E. and Goni, I. B. (1999) Lakes, groundwater and palaeohydrology in the Sahel of NE Nigeria: evidence from hydrogeochemistry. *J. Geol. Soc. London* **156**, 345–355.

Edmunds, W. M., Guendouz, A. H., Mamou, A., Moulla, A., Shand, P. and Zouari, K. (2003) Groundwater evolution in the Continental Intercalaire aquifer of southern Algeria and Tunisia: trace element and isotopic indicators. *Appl. Geochem. Applied Geochemistry*, **18**, 805–822.

Fontes, J -Ch. (1980) Environmental isotopes in groundwater hydrology. In: *Handbook of Environmental Isotope Geochemistry*, Vol. 1, ed. P. Fritz and J -Ch. Fontes, 75–140, Elsevier.

Fontes, J -Ch. andrews, J. N., Edmunds, W. M., Guerre, A. and Travi, Y. (1991) Palaeorecharge by the Niger River (Mali) deduced from groundwater chemistry. *Water Resour. Res.* **27**, 199–214.

Fontes, J -Ch., Gasse, F. and Andrews, J. N. (1993) Climatic conditions of Holocene groundwater recharge in the Sahel zone of Africa. In *Isotopic Techniques in the Study of Past and Current Environmental Changes in the Hydrosphere and the Atmosphere*, 231–248. IAEA.

Gasse, F. (2000) Hydrological changes in the African tropics since the Last Glacial Maximum. *Quaternary Science Rev.* **19**, 189–211.

Gaye, C. B. and Edmunds, W. M. (1996) Intercomparison between physical, geochemical and isotopic methods for estimating groundwater recharge in northwestern Senegal. *Environmental Geology* **27**, 246–251.

Gonfiantini, R., Conrad, G., Fontes, J -Ch., Sauzay, G. and Payne, B. R. (1974) Etude isotopique de la nappe du CI et ses relations avec les autres nappes du Sahara Septentrional. In *Proc. Symposium on Isotope Hydrology, Vienna, Mar. 1974*, 227–241, IAEA.

Goni, I. B., Fellmen, E. and Edmunds, W. M. (2001) Rainfall geochemistry in the Sahel region of northern Nigeria. *Atmospheric. Environ.* **35**, 4331–4339.

Heaton, T. H. E., Talma, A. S. and Vogel, J. C. (1983) Origin and history of nitrate in confined groundwater in the western Kalahari, *J. Hydrol.* **62**, 243–262.

Heaton, T. H. E. (1984) Sources of the nitrate in phreatic groundwater in the western Kalahari. *J. Hydrol.* **67**, 249–259.

Herczeg, A. L. and Edmunds, W. M. (1999) Inorganic ions as tracers. In: *Environmental Tracers in Subsurface Hydrology*, ed. P. G. Cook and A. L. Herczeg, 31–77, Kluwer.

Hingston, F. J. and V. Gailitis. (1976) The geographic variation of salt precipitated over Western Australia. *Australian Journal of Soil Research* **14**, 319–335.

Issar, A. S. (1990) *Water shall Flow from the Rock: Hydrogeology and Climate in the Lands of the Bible*. Springer-Verlag.

Junge, C. E. and Werby, R. T. (1958) The concentration of chloride, sodium, calcium and sulphate in rain water over the United States. *J. Meteorol.* **15**, 417–425.

Kaufmann, R. S., Frape, S. K., McNutt, R. and Eastoe, C. (1993) Chlorine stable isotope distribution of Michigan Basin formation waters. *Appl. Geochem.* **8**, 403–407.

Kinniburgh, D. G. and Miles, D. L. (1983) Extraction and chemical analysis of interstitial water from soils and rocks. *Environmental Science and Technology* **17**, 362.

Lehmann, B. E., Loosli, H. H., Purtschert, R. and Andrews, J. (1995) A comparison of chloride and helium concentrations in deep groundwaters. In *Isotopes in Water Resources Management*, 3–17, IAEA.

Loosli, H. H., Lehmann, B. E. and Smethie, W. M. Jr. (1999) Atmospheric noble gases. In *Environmental Tracers in Subsurface Hydrology*, ed. P. G. Cook and A. L. Herczeg, 349–378, Kluwer.

Love, A. J., Herczeg, A. L., Sampson, L., Cresswell, R. G. and Fifield, L. K. (2000) Sources of chloride and implications for 36Cl dating of old groundwater, southwestern Great Artesian Basin, Australia. *Water Resour. Res.* **36**, 1561–1574.

Ma, J. and Edmunds, W. M. (2006) Groundwater and lake evolution in the Badain Jaran desert ecosystem, Inner Mongolia. *Hydrogeology Journal*, **14**, 1231–1243.

Pearson, F. J. and Swarzenski, W. V. (1974) 14C evidence for the origin of arid region groundwater, northeastern province, Kenya. In *Isotope Techniques in Groundwater Hydrology*, Vol. II, 95–109, IAEA.

Phillips, F. M. (1994) Environmental tracers for water movement in desert soils of the American southwest. *Soil Sci. Soc. Amer. J.* **58**, 15–24.

Phillips, F. M. (1999) Chlorine-36. In *Environmental Tracers in Subsurface Hydrology*, ed. P. G. Cook and A. L. Herczeg, 299–348, Kluwer.

Rozanski, K., Araguas-Araguas, L. and Gonfiantini, R. (1993) Isotopic patterns in modern precipitation. *AGU, Geophysical Monograph* **78**, 1–36.

Servant, M. and Servant-Vildary, S. (1980) L'environnement quaternaire du bassin du Tchad. In *The Sahara and the Nile*, ed. M. A. J. Williams and H. Faure, 133–162, Balkema.

Sonntag, C., Thorweihe, U., Rudolph, J. *et al.* (1980) Isotopic identification of Saharan groundwaters, groundwater formation in the past. *Palaeoecology of Africa* **12**, 159–171.

Stute, M. and Schlosser, P. (1999) Atmospheric noble gases. In *Environmental Tracers in Subsurface Hydrology*, ed. P. G. Cook and A. L. Herczeg, 349–377, Kluwer.

Tardy, Y. (1970) Characterization of the principal weathering types by the geochemistry of waters from some European and African crystalline massifs. *Chemical Geology* **7**, 253–271.

Taupin, J-D., Gallaire, R. and Arnaud, Y. (1997) Analyses isotopiques et chimiques des précipitations sahéliennes de la région de Niamey au Niger: implications climatologiques. In *Hydrochemistry*, ed. N. E. Peters, A. Coudrain-Ribstein, IAHS Publ. **244**, 151–162.

Taupin, J-D., Coudrain-Ribstein, A., Gallaire, R., Zuppi, G. M. and Filly, A. (2000) Rainfall characteristics (δ18O, δ2H, ΔT, ΔH$_r$) in western Africa: Regional scale and influence of irrigated areas. *J. Geophys. Res.* **105**, 11911–11924.

Tyler, S. W., Chapman, J. B., Conrad, S. H. *et al.* (1996) Soil-water flux in the southern Great Basin, United States: Temporal and spatial variations over the last 120 000 years. *Water Resour. Res.* **32**, 1481–1499.

Vengosh, A. and Spivak, A. J. (1999) Boron isotopes in groundwater. In *Environmental Tracers in Subsurface Hydrology*, ed. P. G. Cook and A. L. Herczeg, 479–485, Kluwer.

Winchester, J. W. and Duce, R. A. (1967) The global distribution of iodine, bromide and chloride in marine aerosols. *Naturwissenschaften* **54**, 110–113.

Winograd, I. J. and Robertson, F. N. (1982) Deep oxygenated groundwater: anomaly or common occurrence? *Science* **216**, 1227–1230.

Wood, W. W. (1999) Use and misuse of the chloride mass balance method in estimating groundwater recharge. *Ground Water* **37**, 2–3.

Wood, W. and Sanford, W. E. (1995) Chemical and isotopic methods for quantifying ground-water recharge in regional, semi-arid environment. *Ground Water* **33**, 458–468.

Yechieli, Y. A., Starinsky, A. and Rosental, E. (1992) Evolution of brackish groundwater in a typical arid region: northern Arava Rift Valley, southern Israel. *Appl. Geochem.* **7**, 361–373.

Zimmermann, U., Munnich, K. O. and Roether, W. (1967) Downward movement of soil moisture traced by means of hydrogen isotopes. In *Isotope Techniques in the Hydrologic Cycle*, ed. E. S. Glenn, Am. Geophys. Union, Geophys. Monograph **11**, 28–36.

Zouari, K., Aranyossy, J. F. and Fontes, J -Ch. (1985) Etude isotopique et géochimique des mouvements et de l'évolution des solutions de la zone aérée des sols sous climat semi-aride (sud Tunisien). IAEA-TECDOC-357, 121–143, IAEA.

4 Groundwater flow and transport

J. Carrera and S. A. Mathias

4.1 INTRODUCTION

A model is an entity built to reproduce some aspect of the behaviour of a natural system. In the context of groundwater, aspects to be reproduced may include: groundwater flow (heads, water velocities, etc.); solute transport (concentrations, solute fluxes, etc.); reactive transport (concentrations of chemical species reacting among themselves and with the solid matrix, minerals dissolving or precipitating, etc.); multiphase flow (fractions of water, air, non-aqueous phase liquids, etc.); energy (soil temperature, surface radiation, etc.); and so forth.

Depending on the type of description of reality that one is seeking (qualitative or quantitative), models can be classified as conceptual or mathematical. A conceptual model is a qualitative description of 'some aspect of the behaviour of a natural system'. This description is usually verbal, but may also be accompanied by figures and graphs. In the groundwater flow context, a conceptual model involves defining the origin of water (areas and processes of recharge) and the way it flows through and exits the aquifer. In contrast, a mathematical model is an abstract description (abstract in the sense that it is based on variables, equations and the like) of 'some aspect of the behaviour of a natural system'. However, the motivation of mathematical models is not abstract, but to aid quantification. For example, a mathematical model of groundwater flow should yield the time evolution of heads and fluxes (water movements) at every point in the aquifer.

Both conceptual and mathematical models seek understanding. Some would argue that understanding is not possible without quantification. Conversely, one cannot even think of writing equations without some sort of qualitative understanding. The methods of conceptual modelling are those of conventional hydrogeology (study geology, measure heads and hydraulic parameters, hydrochemistry, etc). The methods of mathematical modelling (discretisation, calibration, etc.) are more specific. Yet it should be clear from the outset that conceptualisation is the first step in modelling and that mathematical modelling helps in building firm conceptual models.

4.1.1 Modelled phenomena

The most basic phenomenon is groundwater flow (Figure 4.1), both because of its intrinsic importance and because it is needed for subsequent processes. In essence, the flow equation expresses two things. First, groundwater movement is normally represented according to Darcy's law. Second, a mass balance must be satisfied in the whole aquifer and in each of its parts. Therefore, the main output from flow models is a mass balance: classified inflows, outflows and storage variations. The output also includes water flows through the aquifer (water fluxes) and heads (water levels in the aquifer). In essence, input data are a thorough description of hydraulic conductivity (and/or transmissivity), storativity, recharge/discharge throughout the model domain, as well as conditions at the model boundaries. Obviously, these data are never available, and the modeller has to use a good deal of ingenuity to generate them. This is where the conceptual model becomes important.

Conservative transport refers to the movement of inert substances dissolved in water. Solutes are affected by advection (displacement of the solute as linked to flowing water) and dispersion (dilution of contaminated water with clean water, which causes the size of the contaminated area to grow while reducing peak concentrations). The main input to a solute transport model is the output of a flow model (water fluxes). Additionally, porosity and dispersivity need to be specified (Figure 4.1). The output is the time evolution and spatial distribution of concentrations. While the amount of additional data needed for solute transport modelling is relatively small, it must be stressed that solute transport is extremely sensitive to variability and errors in water fluxes. A flow model may be good enough for flow results (heads and water balances) but insufficiently detailed to yield water fluxes good enough for

Groundwater Modelling in Arid and Semi-Arid Areas, ed. Howard S. Wheater, Simon A. Mathias and Xin Li. Published by Cambridge University Press.

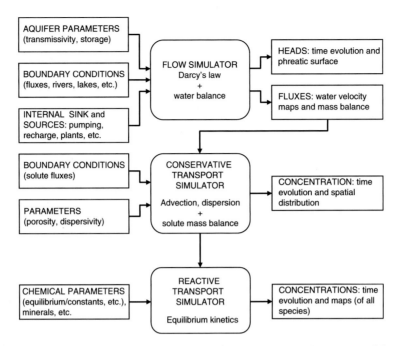

Figure 4.1. A groundwater flow model involves using a flow simulator to take aquifer parameters, boundary conditions and internal sink and sources as inputs and obtain heads and water fluxes as output. Water fluxes are used in conservative transport models, together with porosity, diffusivity and solute mass inflows, to yield the time evolution and spatial distribution of inert tracers.

solute transport. Therefore, modelling solute transport ends up being rather difficult.

Reactive transport refers to the movement of solutes that react among themselves and with the soil phase. Reactions can be of many kinds, ranging from sorption of a contaminant onto a solid surface to redox phenomena controlling the degradation of an organic pollutant. Input for reactive transport modelling includes the output of flow and conservative transport models but also the equilibrium constants of the reactions (usually available from chemistry databases) and the parameters controlling reaction kinetics. However, the most difficult input is the proper identification of relevant chemical processes. Model output includes the concentrations of all chemical species, the reaction rates, etc.

Coupled models refer to models in which different phenomena are affected reciprocally. Density dependent flow is a typical example. Density may depend on solute concentrations. Therefore, density is an output from transport. However, variations in density affect groundwater flow (e.g. dense seawater sinks under light freshwater) which in turn affects solute transport and, hence, density distribution. Other coupled phenomena are the non-isothermal flow of water (coupling flow and energy transport) or the mechanically driven flow of water (coupling flow and mechanical deformation equations).

4.1.2 What are models built for?

In discussing what models can be used for, it is convenient to distinguish between site-specific models and generic models. The former are aimed at describing a specific aquifer, while the latter emphasise processes, regardless of where they take place.

Groundwater management is the ideal use of site-specific models. Management involves deciding where to extract and/or inject water so as to satisfy water needs while ensuring water quality and other constraints. In this context, it is important to point out that a model is essentially a system for accounting for water fluxes and stores (Figure 4.2), in the same way that the accounting system of a company keeps track of money fluxes and reserves. No one would imagine a well-managed company without a proper accounting system. Aquifers will not be managed accurately until they have a model running on real time. Unfortunately, at present, this is still a dream. Because of the difficulties in building and maintaining models and because of legal and practical difficulties to manage aquifers in real time, models are rarely, if ever, used in this fashion.

Instead, models are often used as decision support tools. Building an accurate model is very difficult and time-consuming. As a result, one can rarely expect models to yield exact predictions. However, approximate models are much easier to build. These do not result in precise forecasts but normally allow reasonable assessments of the outcome of different management alternatives. That is, the relative advantages and disadvantages of each alternative can be evaluated and the options be ranked. This is usually all one needs for decision making.

This type of use is frequent in supporting aquifer exploitation policies, that is, for answering questions such as: 'How much

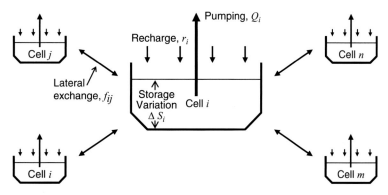

Figure 4.2. A groundwater model is the accounting system of an aquifer. It keeps track of the balance of each section (cells or compartments in the groundwater language) by evaluating exchanges with the outside (pumping, Q_i, recharge, r_i, etc.) and with the adjacent sections (f_{ij}). The difference between inflow and outflow is equal to the variation in reserves (storage variation, ΔS_i). A well-managed company needs an accounting system, and so does an aquifer.

water can be extracted?', 'Where should one pump to minimise environmental impact?' In fact, a large body of literature is devoted to this kind of question in an optimal fashion (see Gorelick, 1983, to start a review of this topic).

Site-specific models are most frequently used, however, as a tool to support aquifer characterisation efforts. This is somewhat ironic because a model is an essentially quantitative tool, while site characterisation is rather qualitative. Yet experience dictates that modelling is the only way to consistently integrate the kind of data available in site characterisation. These data are very diverse. They may range from geologic maps to isotope concentrations. One can use vastly different models to verbally explain all observations. Quantitative consistency is not so easy to check and requires the use of a model. Because of the difficulties in fully describing all data, this kind of model use is rarely described in the scientific literature.

Models can also be used in a generic fashion as teaching or research tools to gain understanding in physicochemical phenomena. In these cases, they do not aim at representing a specific aquifer, but at evaluating the role of some processes under idealised conditions. A classic example of this type of use is the analysis of flow in regional basins (Freeze and Witherspoon, 1966). Models are also used in this fashion to explain geological processes (Garven and Freeze, 1984; Ayora *et al.*, 1998; Steefel *et al.*, 2005).

4.1.3 Scope of the chapter

Modelling involves a large number of issues, ranging from understanding site geology to numerical methods and the way to use these in order to represent the studied aquifer. It is clear that one chapter in a book cannot cover in detail all these issues. This chapter has been arranged to cover the basics of the most important, while directing the reader to specialised literature for details. Because of the breadth of materials,

descriptions are restricted to flow problems (i.e. transport and other phenomena are not discussed further except for the presentation of the basic solute transport equations, below). The chapter starts by describing the flow phenomenon in terms of equations and basic numerical methods. This is followed by a discussion on the general procedure of building models using numerical codes. Since heterogeneity is such an ubiquitous problem in hydrogeology, the chapter dedicates a large section describing many of the basic tools available to deal with spatial and temporal variability. From here, a formalisation of the calibration procedure is made. Finally, the chapter concludes with a discussion on methods for uncertainty assessment.

4.2 GOVERNING EQUATIONS OF FLOW AND TRANSPORT

4.2.1 Mass conservation

The general fluid mass equation that is usually referred to as the groundwater flow equation is (see Bear, 1979, for details)

$$\frac{\partial \rho \varepsilon S_w}{\partial t} - \nabla \cdot \left[\left(\frac{\mathbf{k}k_r\rho}{\mu} \right) \cdot (\nabla p + \rho \mathbf{g}) \right] = Q_p \tag{4.1}$$

where S_w [-] is water saturation, ε [-] is porosity, p [ML^{-1}T^{-2}] is fluid pressure, t [T] is time, $\mathbf{k}$ [L^2] is the permeability tensor, k_r [-] is the relative permeability for unsaturated flow, μ [ML^{-1}T^{-1}] is the fluid viscosity, $\mathbf{g}$ [LT^{-2}] is the gravity vector, Q_p [ML^{-3}T^{-1}] is a fluid mass source and ρ [ML^{-3}] is the fluid density. We note that this applies to the great majority of problems. However, in some situations equation (4.1) must be substantially modified, particularly when dual-permeability effects and/or inertial/turbulent flow effects are prevalent (e.g. in karst systems or in the vicinity of a well) (see Wu, 2002, and Mathias *et al.*, 2008, for further discussion).

The fluid flux $\mathbf{q}$ [LT^{-1}] (often referred to as the Darcy flux) is given by Darcy's law:

$$\mathbf{q} = -\left(\frac{\mathbf{k}k_r}{\mu}\right) \cdot (\nabla p + \rho \mathbf{g}) \qquad (4.2)$$

The solute mass balance equation referred to as the transport equation is

$$[\varepsilon S_w \rho + (1-\varepsilon)\rho_s \kappa_1]\frac{\partial C}{\partial t} + \rho \mathbf{q} \cdot \nabla C - \nabla \cdot [(\varepsilon S_w D_m + \mathbf{D}) \cdot \nabla C]$$
$$= Q_p(C^* - C) + \varepsilon S_w \rho(\gamma_1^w C + \gamma_0^w) + (1-\varepsilon)\rho_s(\gamma_1^s C_s + \gamma_0^s) \qquad (4.3)$$

where κ_1 [-] is a sorption coefficient defined in terms of the selected equilibrium sorption isotherm, ρ_s [ML^{-3}] is the density of the solid grains, C [MM^{-1}] is the solute mass fraction, D_m [L^{2}T^{-1}] is the coefficient of molecular diffusion for the porous medium fluid, $\mathbf{D}$ [L^{2}T^{-1}] is the dispersion tensor, C^* [MM^{-1}] is the solute mass fraction of a fluid source, γ_1^w [T^{-1}] is the first-order solute production rate, γ_1^s [T^{-1}] is the first-order sorbate production rate, γ_0^w [M$_{solute}$M$_{fluid}^{-1}$T^{-1}] is the zero-order solute production rate, and γ_0^s [M$_{solute}$M$_{fluid}^{-1}$T^{-1}] is the zero-order sorbate production rate.

4.2.2 Reduction to the two-dimensional groundwater equation

Chapter 7 discusses issues associated with variable density flow in detail. For the remainder of this chapter it is assumed that the solute concentration is sufficiently small so as not to significantly affect the fluid density. For a fluid of constant density, equation (4.1) reduces to

$$\frac{\partial \varepsilon S_w}{\partial t} - \nabla \cdot \left[\left(\frac{\mathbf{k}k_r}{\mu}\right) \cdot (\nabla p + \rho \mathbf{g})\right] = \frac{Q_p}{\rho} \qquad (4.4)$$

In this case it is often helpful to write the equation in terms of hydraulic head, $h = p/\rho g + z$ [L] where z [L] is the elevation. Working with hydraulic head is advantageous for several reasons. Hydraulic head is easily measurable because it is equal to the elevation of water in piezometers. Head maps are easy to read because water flows perpendicular to head contours. Furthermore, it allows Darcy's law (equation (4.2)) to be written in the simpler form

$$\mathbf{q} = -\mathbf{K} \cdot \nabla h \qquad (4.5)$$

where $\mathbf{K} = \rho g \mathbf{k} k_r/\mu$ is referred to as the hydraulic conductivity tensor. The flow equation (4.4) becomes

$$\frac{\partial \varepsilon S_w}{\partial t} - \nabla \cdot [\mathbf{K} \cdot \nabla h] = \frac{Q_p}{\rho} \qquad (4.6)$$

Indeed, most groundwater models ignore the unsaturated zone and assume that hydraulic head is constant with elevation (the Dupuit assumption). Further simplification can be achieved by taking $\mathbf{K}$ to be constant along the vertical, assuming the aquifer to be flat and head to be referred to the aquifer base (a set of frequent though rarely justified assumptions), such that equation (4.6) reduces to

$$S_y \frac{\partial h}{\partial t} - \frac{\partial}{\partial x}\left(K_x h \frac{\partial h}{\partial x}\right) - \frac{\partial}{\partial y}\left(K_y h \frac{\partial h}{\partial y}\right) = r \qquad (4.7)$$

where K_x [LT^{-1}] and K_y [LT^{-1}] are representative, depth-averaged hydraulic conductivities in the horizontal x [L] and y [L] directions, S_y [-] is referred to as the specific yield (the volume of water released per unit area per unit drop in head) and r [LT^{-1}] is the aquifer recharge per unit area.

For confined aquifers, or for free aquifers where $\mathbf{K}$ at the bottom of the aquifer is much larger than at the phreatic surface, equation (4.7) reduces further to

$$S \frac{\partial h}{\partial t} - \frac{\partial}{\partial x}\left(T_x \frac{\partial h}{\partial x}\right) - \frac{\partial}{\partial y}\left(T_y \frac{\partial h}{\partial y}\right) = r \qquad (4.8)$$

where $T_x = m K_x$ [L^{2}T^{-1}] and $T_y = m K_y$ [L^{2}T^{-1}] are transmissivities in the x and y direction and $S = S_y$ for free aquifers or $S = m S_s$ [-], which is referred to as the storativity, where m is the aquifer thickness and S_s [L^{-1}] is the specific storage coefficient associated with the aquifer compressibility. In this context, r is a recharge associated with some form of leakage through the overlying aquitard. Notice also that T is, in practice, what hydrogeologists measure from pumping tests (it is arguably the most important parameter for field hydrogeologists). That is, transmissivity is never derived as the product of a hydraulic conductivity times an aquifer thickness; it is only written like that for consistency with the derivation procedure.

4.2.3 Solution by finite difference

Analytical solution of equations (4.7) and (4.8) is only possible for highly idealised situations. For most practical purposes it is necessary to apply some form of numerical approximation. Arguably one of the most commonly used groundwater modelling packages is the USGS MODFLOW code (McDonald and Harbaugh, 1988) whereby equations (4.7) and (4.8) are solved using finite differences (Richtmeyer and Morton, 1957). Other techniques often used in groundwater modelling include finite volumes (e.g. Pruess *et al.*, 1999) and finite elements (e.g. Diersch, 2005).

As an illustrative example, let us consider the finite difference approximation of equation (4.8). The starting point is to consider the two Taylor expansions:

$$f(x + \Delta x) = f(x) + \Delta x \frac{df}{dx} + \frac{\Delta x^2}{2!}\frac{d^2f}{dx^2} + \frac{\Delta x^3}{3!}\frac{d^3f}{dx^3} + \cdots \qquad (4.9)$$

$$f(x - \Delta x) = f(x) - \Delta x \frac{df}{dx} + \frac{\Delta x^2}{2!}\frac{d^2f}{dx^2} - \frac{\Delta x^3}{3!}\frac{d^3f}{dx^3} + \cdots \qquad (4.10)$$

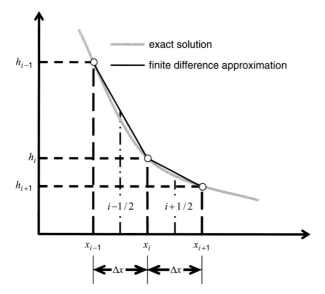

Figure 4.3. Finite differences consists essentially of approximating the tangent (slope) of the piezometric surface by its secant.

which rearrange to obtain the forward and backward finite difference approximations (Figure 4.3):

$$\frac{df}{dx} = \frac{f(x + \Delta x) - f(x)}{\Delta x} + O(\Delta x) \tag{4.11}$$

$$\frac{df}{dx} = \frac{f(x) - f(x - \Delta x)}{\Delta x} + O(\Delta x) \tag{4.12}$$

Now we discretise time and space into the set of nodes x_i, y_j and t_k where $i = 1 \ldots N_i$, $j = 1 \ldots N_j$ and $k = 1 \ldots N_k$ and N_i, N_j, N_k are the number of nodes in the x, y and t directions respectively. For simplicity let us also assume that the nodes are equally spaced in their respective dimensions by Δx [L], Δy [L] and Δt [T] (although it is trivial to relax this assumption). It follows that $h(x_i, y_j, t_k)$ can be approximated by $h_{i,j,k}$ found from

$$S_{i,j}\left(\frac{h_{i,j,k} - h_{i,j,k-1}}{\Delta t}\right) - \frac{1}{\Delta x}\left[T_{x,i+1/2,j}\left(\frac{h_{i+1,j,k} - h_{i,j,k}}{\Delta x}\right) - T_{x,i-1/2,j}\left(\frac{h_{i,j,k} - h_{i-1,j,k}}{\Delta x}\right)\right]$$
$$- \frac{1}{\Delta y}\left[T_{y,i,j+1/2}\left(\frac{h_{i+1,k} - h_{i,j,k}}{\Delta y}\right) - T_{y,i,j-1/2}\left(\frac{h_{i,j,k} - h_{i-1,k}}{\Delta y}\right)\right] = r_{i,j,k} \tag{4.13}$$

where

$$\begin{aligned}
T_{x,i+1/2,j} &= 2/(1/T_{x,i+1,j} + 1/T_{x,i,j}) \\
T_{x,i-1/2,j} &= 2/(1/T_{x,i-1,j} + 1/T_{x,i,j}) \\
T_{y,i,j+1/2} &= 2/(1/T_{y,i,j+1} + 1/T_{y,i,j}) \\
T_{y,i,j-1/2} &= 2/(1/T_{y,i,j-1} + 1/T_{y,i,j})
\end{aligned} \tag{4.14}$$

and $S_{i,j}$, $T_{x,i,j}$, $T_{y,i,j}$ and $r_{i,j,k}$ are approximations to the continuous fields of S, T_x, T_y and r respectively.

It is interesting to consider the physical meaning of equation (4.13). For this purpose, multiply both sides by $\Delta x \Delta y$ (the area of cell i) such that equation (4.13) becomes

$$\Delta S_{i,j} = f_{i+1 \to i,j} + f_{i-1 \to i,j} + f_{i,j+1 \to j} + f_{i,j-1 \to j} + Q_{i,j} \tag{4.15}$$

where $\Delta S_{i,j} = S\Delta x \Delta y(h_{i,j,k} - h_{i,j,k-1})/\Delta t$ is the variation in water volume stored in the cell per unit time (recall that S is the variation in water volume stored per unit surface and unit change in head). The f terms represent the rates of water volume exchanged between the i, j and adjacent cells. $Q_{i,j} = \Delta x \Delta y r_{i,j}$ is the flow rate of recharge. The summary of this discussion is that the finite difference approximation expresses the discrete (cell-wise) version of the two principles on which the flow equation is based (Darcy's law and mass conservation).

Reverting back to before, equation (4.13) can be written for all cells, which allows $h_{i,j,k}$ to be found from the system of equations:

$$\begin{bmatrix} C_{1,1} & D_{1,1} & & E_{1,1} & & & \\ B_{2,1} & C_{2,1} & D_{2,1} & & E_{2,1} & & \\ & \ddots & \ddots & \ddots & & \ddots & \\ & & B_{N,1} & C_{N,1} & 0 & & E_{N,1} \\ A_{1,2} & & & 0 & C_{1,2} & D_{1,2} & & E_{1,2} \\ & A_{2,2} & & & B_{2,2} & C_{2,2} & D_{2,2} & & E_{2,2} \\ & & \ddots & & & \ddots & \ddots & \ddots & & \ddots \\ & & & A_{N,2} & & & B_{N,2} & C_{N,2} & 0 & \\ & & & & \ddots & & & \ddots & \ddots & \ddots \\ & & & & & A_{N,N_j} & & & B_{N,N_j} & C_{N,N_j} \end{bmatrix} \begin{bmatrix} h_{1,1,k+1} \\ h_{2,1,k+1} \\ \vdots \\ h_{N,1,k+1} \\ h_{1,2,k+1} \\ h_{2,2,k+1} \\ \vdots \\ h_{N,2,k+1} \\ \vdots \\ h_{N,N_j,k+1} \end{bmatrix} = \begin{bmatrix} F_{1,1}h_{1,1,k} + r_{1,1,k+1} \\ F_{2,1}h_{2,1,k} + r_{2,1,k+1} \\ \vdots \\ F_{N,1}h_{N,1,k} + r_{N,1,k+1} \\ F_{1,2}h_{1,2,k} + r_{1,2,k+1} \\ F_{2,2}h_{2,2,k} + r_{2,2,k+1} \\ \vdots \\ F_{N,2}h_{N,2,k} + r_{N,2,k+1} \\ \vdots \\ F_{N,N_j}h_{N,N_j,k} + r_{N,N_j,k+1} \end{bmatrix} \tag{4.16}$$

where

$$\begin{aligned}
A_{i,j} &= \left(\frac{T_{y,i,j-1} + T_{y,i,j}}{2\Delta y^2}\right), B_{i,j} = \left(\frac{T_{x,i-1,j} + T_{x,i,j}}{2\Delta y^2}\right) \\
C_{i,j} &= \frac{S_{i,j}}{\Delta t} + \left(\frac{T_{x,i+1,j} - T_{x,i-1,j}}{2\Delta x^2}\right) + \left(\frac{T_{y,i,j+1} - T_{y,i,j-1}}{2\Delta y^2}\right) \\
D_{i,j} &= -\left(\frac{T_{x,i+1,j} + T_{x,i,j}}{2\Delta y^2}\right), E_{i,j} = -\left(\frac{T_{y,i,j+1} + T_{y,i,j}}{2\Delta y^2}\right) \\
F_{i,j} &= \frac{S_{i,j}}{\Delta t}
\end{aligned} \tag{4.17}$$

Using vector notation, equation (4.16) can be written as

$$\mathbf{Bh}_{k+1} = \mathbf{b} \tag{4.18}$$

where $\mathbf{B}$ is the coefficients matrix, $\mathbf{h}$ the vector of heads at all nodes, and $\mathbf{b}$ the right-hand side of equation (4.16) such that $\mathbf{b} = \mathbf{f}^T \mathbf{Ih}_k + \mathbf{r}_{k+1}$ where $\mathbf{f}$ is the vector of $S_{i,j}/\Delta t$, $\mathbf{I}$ denotes the identity matrix and $\mathbf{r}$ is the vector of recharge at all nodes. The subscript denotes the time-step. Actual solution of this system of equations is achieved sequentially in time. Starting from initial heads, one builds $\mathbf{b}_0$ and uses equation (4.16) to solve for $\mathbf{h}_1$ (vector of heads at time $k - 1$). The procedure is repeated as shown in Figure 4.4.

Equation (4.16) was derived for finite differences, but similar equations would have been obtained if other numerical methods had been used. Different numerical methods (finite difference, finite elements, finite volumes, boundary elements, etc.) generally lead to different spatial discretisation schemes (see Figure 4.7) and different formulations of the $\mathbf{B}$ matrix. However, all methods share the fact that equation (4.16) essentially expresses mass conservation and Darcy's law. Moreover, all of them require an algorithm similar to that of Figure 4.4.

The transport equation, unlike the flow equation, is very difficult to solve. As a result, numerous methods have been

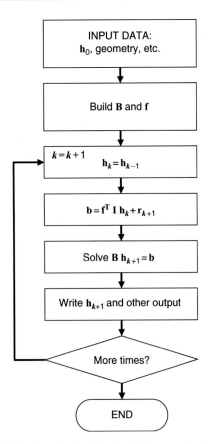

Figure 4.4. Basic steps involved in simulating groundwater flow. Heads $\mathbf{h}^{k+1}$ are computed by solving the equation $\mathbf{B}\,\mathbf{h}^{k+1} = \mathbf{b}$. They may be written as output for plotting. They are used as initial head for the next time increment. These steps (time loop) are repeated sequentially until the last time is reached.

developed to overcome difficulties. The most widely used in groundwater include the above methods, but modified by upwinding or other techniques (Heinrich *et al.*, 1977; Kipp, 1987), the 'method of characteristics' (Konikow and Bredehoeft, 1978), the 'modified method of characteristics' (Neuman, 1981) and the 'Eulerian–Lagrangian localised adjoint method' (Celia *et al.*, 1990) among others. However, while a specific method can be found adequate in most cases, no method is universally accepted as best.

4.3 THE MODELLING PROCEDURE

The procedure to build a model is outlined in Figure 4.5. First, one defines a conceptual model (i.e. zones of recharge, boundaries of aquifers, etc.). Second, one discretises the model domain into a finite element or finite difference grid. This can be entered as input data for a simulation code. Unfortunately, output data will rarely fit the observed aquifer heads and concentrations. This is the motivation for calibration, that is, the modification

of model parameters to ensure that model output is indeed similar to what has been observed in reality. The model thus calibrated can be considered a 'representation of the natural system' and used for management or simulation purposes.

The above procedure is formally described in Figure 4.6. This section is devoted to discussing in detail the different modelling steps (e.g. Anderson and Woessner, 1992; Carrera *et al.*, 1993). In practice, the effort behind each of these tasks is sensitive to the objectives of the modelling study at hand. However, for building models that describe reality in sufficient detail for predictive purposes, some generic issues are relevant.

4.3.1 Conceptualisation

Modelling starts by defining which processes are relevant and how they are to represented within the model. Definition of the relevant processes is termed 'process identification'. Process identification can be difficult for transport problems, where it may not be easy to define beforehand which physical and chemical processes are affecting the solute (does it sorb? does it diffuse into immobile zones? does it react with other solutes? does it degrade? etc.). Groundwater flow is, in principle, easier. Still, a number of questions must be answered (is the aquifer unconfined? does transmissivity change with saturated thickness? does the aquifer interact with the river? etc.). To answer these questions, the modeller must make simplifications and assumptions. Many of these can be revised later.

Model structure identification refers to the definition of parameter variability, boundary conditions, etc. In a somewhat narrower but more systematic sense, model structure identification implies expressing the model in terms of a finite number of unknowns, the 'model parameters'. Parameters controlling the above processes are variable in space. In some cases, they also vary in time or depend on heads and/or concentrations. As discussed earlier, data are scarce so that such variability cannot be expressed accurately. Therefore, the modeller is also forced to make numerous simplifications to express the patterns of parameter variations, boundary conditions, etc. These assumptions are reflected in what is denoted as the 'model structure'.

The conceptualisation step of any modelling effort is somewhat subjective and dependent on the modeller's ingenuity, experience, scientific background and way of looking at the data. Selection of the physicochemical processes to be included in the model is rarely the most difficult issue. The most important processes affecting the movement of water and solutes underground (advection, dispersion, sorption, etc.) are relatively well known. Ignoring a relevant process will only be caused by misjudgements and should be pointed out by reviewers, which illustrates why reviewing by others is important. Difficulties arise when trying to characterise those processes and, more specifically, the spatial variability of controlling parameters.

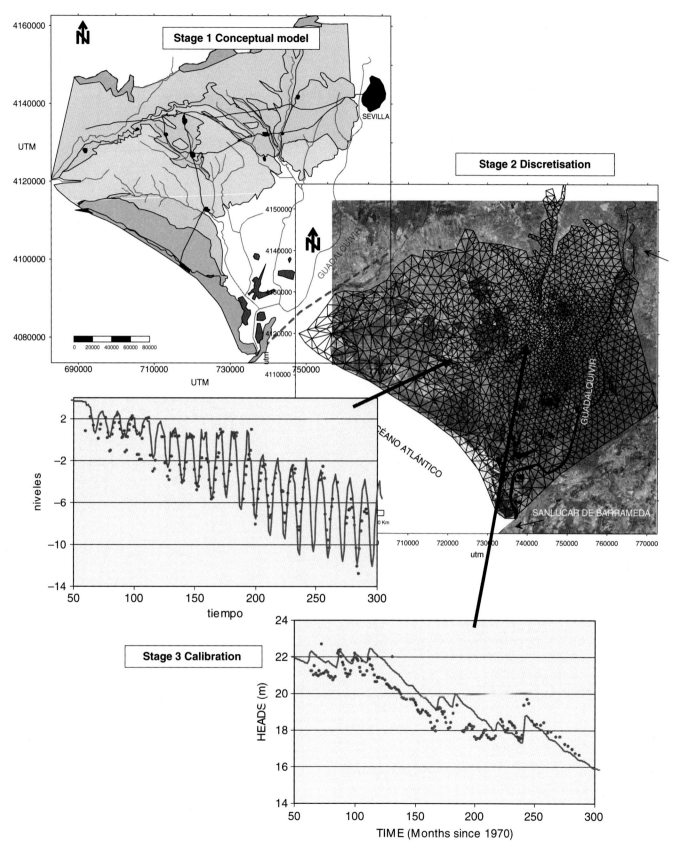

Figure 4.5. Building a model involves three basic steps: conceptualisation, discretisation and calibration. Example from the Almonte-Marismas aquifer (Castro *et al.*, 1999).

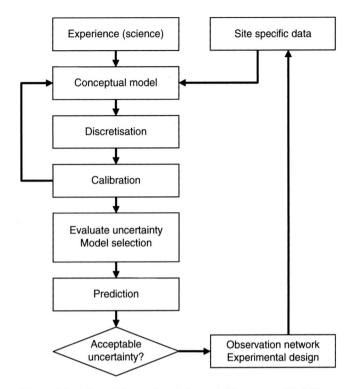

Figure 4.6. A formal description of the modelling process. Modelling starts with an understanding of the natural system (conceptual model), which is based on experience about such kinds of system (science) and on data from the site. Writing the conceptual model in a manner adequate for computer solution requires discretisation. The resulting model is still dependent on many parameters that are uncertain. During calibration, these parameters are adjusted so that model outputs are close to measurements. Model predictions may be uncertain because so are the fitted parameters or because different models are consistent with observations. If uncertainty is unacceptably high, one should perform additional measurements or experiments and redo the whole process.

In spite of the large amount of data usually available, their qualitative nature prevents a detailed definition of the conceptual model. Thus, more than one description of the system may result from the conceptualisation step. Selecting one conceptual model among several alternatives is sometimes performed during calibration, as discussed later.

4.3.2 Discretisation

Strictly speaking, discretisation consists of substituting a continuum by a discrete system (see Figure 4.7). However, we are extending this term here to describe the whole process of going from mathematical equations, derived from the conceptual model, to numerical expressions that can be solved by a computer. Closely related is the issue of verification, which refers to ensuring that a code solves accurately the equations that it is

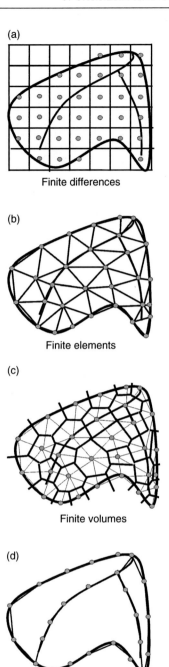

Figure 4.7. The most widely used methods of discretisation are (a) finite differences, which consists of subdividing the model domain into regular rectangles, and (b) finite elements, which is based on dividing the aquifer region into elements of arbitrary shape (often triangles). (c) Finite volumes, also called integrated finite differences, divides the region into polygons. (d) The boundary element method is very convenient, when applicable, because it only requires discretising boundaries (both internal and external).

claimed to solve. As such, verification is a code-dependent concept. However, using a verified code is not sufficient for mathematical correctness. One should also make sure that time and space discretisation is adequate for the problem being addressed. Moreover, numerical implementation of a conceptual model is not always straightforward. International code comparison projects, such as INTRANCOIN and HYDROCOIN, have shown the need for sound conceptual models and independent checks of calculation results. Even well-posed mathematical problems lead to widely different solutions when solved by different people, because of slight variations on the solution methodology or misinterpretations in the formulation (NEA-SKI, 1990).

The main concern during discretisation is accuracy. In this sense, it is not conceptually difficult, although it can be complex. Accuracy is not restricted to numerical errors (differences between numerical and exact solutions of the involved equations) but also refers to the precision with which the structure of spatial variability reproduces the natural system.

4.3.3 Calibration and error analysis

The choice of numerical values for model parameters is made during calibration, which consists of finding those values which grant a good reproduction of head and concentration data (see Figure 4.5) and are consistent with prior independent information.

Calibration is rarely straightforward. Data come from various sources, with varying degrees of accuracy and levels of representativeness. Some parameters can be measured directly in the field, but such measurements are usually scarce and prone to error. Furthermore, since measurements are most often performed on scales and under conditions different from those required for modelling purposes, they tend to be both numerically and conceptually different from model parameters. The most dramatic example of this is dispersivity, whose representative value increases with the scale of measurement (Gelhar, 1986). Therefore dispersivities derived from tracer tests cannot be used directly in a large-scale model. As a result, model parameters are calibrated by ensuring that simulated heads and concentrations are close to the corresponding field measurements.

Calibration can be tedious and time-consuming because many combinations of parameters have to be evaluated, making it prone to be incomplete. This, coupled to difficulties in taking into account the reliability of different pieces of information, makes it very hard to evaluate the quality of results. Therefore, it is not surprising that significant efforts have been devoted to the development of automatic calibration procedures (Carrera and Neuman, 1986a, 1986b, 1986c; Loaiciga and Marino, 1987; Hill, 1992).

4.3.4 Model selection

The first step in any modelling effort involves constructing a conceptual model, describing it by means of appropriate governing equations, and translating the latter into a computer code. Unfortunately, the result is often ambiguous and more than one model may be plausible *a priori*. Model selection involves the process of choosing between alternative model forms. Methods for model selection can be classified into three broad categories. The first category is based on a comparative analysis of residuals (differences between measured and computed system responses) using objective as well as subjective criteria. The second category is denoted parameter assessment and involves evaluating whether or not computed parameters can be considered as 'reasonable'. The third category relies on theoretical measures of model validity known as 'identification criteria'. In practice, all three categories will be needed; residual analysis and parameter assessment suggest ways to modify an existing model and the resulting improvement in model performance is evaluated on the basis of identification criteria. If the modified model is judged an improvement over the previous model, the former is accepted and the latter discarded.

The most widely used tool of model identification is residual analysis. In the groundwater context, the spatial and frequency distributions of head and concentration residuals are very useful in pointing towards aspects of the model that need to be modified. For example, a long tail in the breakthrough curve not properly simulated by a single porosity model may point to a need for incorporating matrix diffusion or a similar mechanism. These modifications should, whenever possible, be guided by independent information. Qualitative data such as lithology, geological structure, geomorphology and hydrochemistry are often useful for this purpose. A particular behavioural pattern of the residuals may be the result of varied causes that are often difficult to isolate. Spatial and/or temporal correlation among residuals may be a consequence of improper conceptualisation, but it may be a consequence of measurement or numerical errors. Simplifications in simulating the stresses exerted over the system are always made and these lead to correlation among residuals. Distinguishing between correlations caused by improper conceptualisation and by measurement errors is not an easy matter and can make analysis of residuals a rather limited tool for model selection.

An expedite way of evaluating a model concept is based on assessing whether or not the parameters representing physico-chemical properties can be considered 'reasonable'; that is, whether or not their values make sense and/or are consistent with those obtained elsewhere. Meaningless parameters can be a consequence of either poor conceptualisation or instability. If a relevant process is ignored during conceptualisation, the effect of such a process may be reproduced by some other parameter. For example, the effect of sorption is to keep part of the solute attached to the solid phase, hence retarding the movement of the solute mass. In linear instantaneous sorption, this effect cannot be distinguished from standard storage in the pores. Therefore, if one needs an absent porosity (e.g. larger than 1) to fit observation, one should consider the possibility of including sorption in the model. However, despite this example, parameter assessment tends to be more useful for ruling out model concepts rather than for guiding how to modify an inadequate model. Residual analysis is usually more helpful for this purpose.

Instability may also lead to unreasonable parameter estimates during automatic calibration, despite the validity of the conceptual model. When the number of data or their information content is low, small perturbations in the measurement or deviations in the model may lead to drastically different parameter estimates. When this happens, the model may obtain equally good fits with widely different parameter sets. Thus, one may converge to a senseless parameter set while missing other perfectly meaningful sets. This type of behaviour can be easily identified by means of a thorough error analysis and corrected by fixing the values of one or several parameters (Carrera and Neuman, 1986a, 1986b, 1986c).

4.3.5   Predictions and uncertainty

Formulation of predictions involves a conceptualisation of its own. Quite often, the stresses whose response is to be predicted lead to significant changes in the natural system, so that the structure used for calibration is no longer valid for prediction. Changes in the hydrochemical conditions or in the flow geometry may have to be incorporated into the model. While numerical models can be used for network design or as investigation tools, most models are built in order to study the response of the medium to various scenario alternatives. Therefore, uncertainties on future natural and human-induced stresses also cause model predictions to be uncertain. Finally, even if future conditions and the conceptual model are exactly known, errors in model parameters will still cause errors in the predictions. In summary, three types of prediction uncertainties can be identified: conceptual model uncertainties; stress uncertainties; and parameter uncertainties.

The first group includes two types of problems. One is related to model selection during calibration; that is, more than one conceptual model may have been properly calibrated and data may not suffice to distinguish which one is the closest to reality. It is clear that such indeterminacy should be carried into the prediction stage because both models may lead to widely different results under future conditions. The second type of problem arises from improper extension of calibration to prediction conditions from not taking into account changes in the natural system or in the scale of the problem. The only way of dealing with this problem is to carefully evaluate whether or not the assumptions on which the calibration was based are still valid under future conditions. Indeed, model uncertainties can be very large.

The second type of uncertainties, those associated with future stresses, is often dismissed as not being part of the modelling procedure. While future stresses may affect the validity of the model, they are external to it. In any case, this type of uncertainty is evaluated by carrying out simulations under a number of alternative scenarios, whose definition is an important subject in itself.

The last set of prediction uncertainties is the one associated with parameter uncertainties, which can be quantified relatively well. This is discussed in detail later in the chapter.

Most uncertainties in groundwater modelling are linked to the spatial and, sometimes, temporal variability. One may argue that spatial variability is the most notable methodological singularity of groundwater hydrology in general, and of groundwater modelling in particular. The next section is entirely devoted to the handling of spatial variability.

4.4   HANDLING SPATIAL VARIABILITY

Variability is an essential feature of some hydrological variables. Permeability, porosity, soil texture, etc. are highly variable in space. They are so variable that one cannot realistically aim at ever knowing them accurately. Increased sophistication and effectiveness of measurement devices causes an increasing level of resolution. Still, the apparently fractal nature of all these phenomena and parameters means that there will always be some level of unresolved variability.

Variability is important not only because of the uncertainty it brings about, but also because large-scale behaviour of a spatially variable phenomenon may be significantly different from the small-scale behaviour. Changes in scale may: (1) lead to changes in the effective parameters; (2) cause the governing equations to change; or (3) cause new processes to emerge. This section is motivated by the recognition that spatial and temporal variability is both very important and impossible to describe accurately.

4.4.1   Importance of geology

Aquifers are the host of groundwater. They are made up of geological materials and they are heterogeneous. Heterogeneity means 'from different origins' and that is squarely what geology is supposed to aim at: understanding the history of the Earth. In our context this is important because: (1) different geological materials display widely varying hydraulic properties; (2) variability is hard to predict and might look random, but it is not – it is controlled by geological history; and (3) variability controls dramatically the behaviour of groundwater.

It may be argued that, of all parameters describing physical properties, hydraulic conductivity is the one spanning most orders of magnitude (Freeze and Cherry, 1979). Measured hydraulic conductivities cover some 14 orders of magnitude.

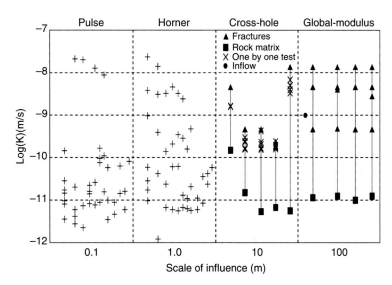

Figure 4.8. Scale dependence of hydraulic conductivity at the Grimsel test site (Martinez-Landa and Carrera, 2005). Whenever highly conductive zones are well connected, one should expect large scale conductivity to be much larger than local conductivity. This example is somewhat extreme because it refers to a crystalline rock, where fractures impose a large connectivity.

Worse, variability occurs at all scales. A permeable aquifer may be adjacent to a low permeability formation. It is frequent to observe clay layers in generally sandy formations, and vice versa. A granite massif can be crossed by highly conductive shear faults that render the medium quite permeable, even if such faults display low transmissivity areas.

Addressing such variability can be done in two ways: (1) attempting to understand the causes of variability (i.e. trying to understand site geology); and (2) concentrating on assessing the effect of small-scale variability on large-scale behaviour. In practice, field hydrogeologists tend to adopt the first avenue, while academicians tend to follow the second. Theoretically, they are not exclusive. One may envision rather accurate descriptions of large-scale geological trends, which cannot be predictive on the small scale, where 'theoretical' results would be more productive. Unfortunately, such blending has not yet occurred. On the other hand, it is believed that theoretical results are maturing at a rate that might make them soon applicable for practitioners. What is clear is that spatial variability of hydraulic parameters cannot be ignored.

Predictions based on the equations presented in Section 4.2 are faulty when variability is ignored. One of the most frequently observed sources of distress is the so-called 'scale effect'. The term refers to the scale dependence of representative parameters, as illustrated by Figure 4.8. Hydraulic conductivities derived from small-scale (say 10 cm) tests average around 10^{-11} m/s, they average around 10^{-10} m/s when working at the meter scale and reach values around 10^{-9} m/s at the 100 m scale. The problem is evident for modelling purposes. Measurements made at one scale may not be appropriate when modelling at a different scale. Explanations for this effect and ways to address it are given in Section 4.4.3.

The problem becomes even more severe for transport problems (Carrera, 1993; Carrera et al., 1998). Dispersivity grows linearly with spatial scale. Porosity appears to depend on flow direction and residence time. Addressing spatial variability is not a luxury; it is a necessity.

In the following, what one could term the 'academic' approach to spatial variability is described. While its practical applicability is limited, it sets the 'right frame of mind' for addressing groundwater problems. Still, readers must bear in mind that geology is the ultimate cause of variability. Therefore, it is required for proper modelling. That we do not enter into it reflects that: (1) geological methods are very site dependent and hard to synthesise; and (2) we hope that geological and stochastic methods will eventually blend.

4.4.2 Geostatistics

In the early days, spatial variability was largely ignored due to the inability of most analytical solutions for groundwater flow equations to handle anything other than a homogeneous system. Serious studies on the spatial variability of aquifer properties came about in the early 1960s alongside the development of the first numerical models, which could easily accommodate spatially varying parameters, recharge and boundary conditions (Stallman, 1956).

The initial approach was to assume that a pumping test derived transmissivity was a local average of the 'real' parameter distribution. A model element that contained a well, which possessed

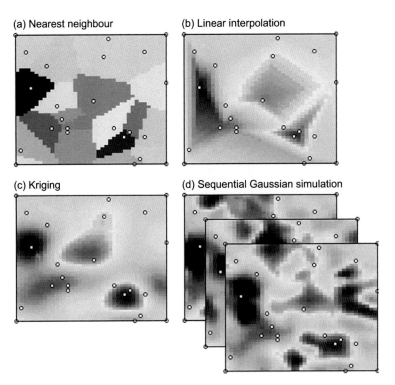

Figure 4.9. Methods to handle spatial variability. Interpolation methods, such as (a), (b) and (c), yield relatively smooth fields, with variability controlled by measurements (i.e. only measured values and locations). Simulation methods such as (d) produce fields whose variability is also controlled by a user-defined statistical structure.

a pumping test result, was assigned that corresponding transmissivity. Model elements that contained no pumping test data were then prescribed parameters on the basis of some interpolation scheme (nearest neighbour, see Figure 4.9a; linear interpolation, see Figure 4.9b; hand contouring, etc.) involving adjacent pumping test results.

A formalism for describing spatial variability was clearly needed. It was brought by Matheron (1971), who developed the concept of regionalised variables. These are functions displaying the same variability as aquifer properties (e.g. transmissivity) but whose actual values are random. While the concept is somewhat abstract, it allows the development of practical algorithms. Below we describe the basic ones: (1) kriging, which allows one to estimate the optimal value of a property (typically the logarithm of hydraulic conductivity), given measurements at a number of points; and (2) conditional simulation, which allows one to generate fields with desired spatial variability patterns, while honouring existing measurements.

KRIGING

A generation of models employed kriging to obtain spatially variable parameter fields (de Marsily *et al.*, 2005). A key assumption is that of statistical homogeneity. The idea is that values of a given parameter, $Z(\mathbf{x}_0)$, can be estimated from

a weighted average of all the known observations in that vicinity, $Z(\mathbf{x}_i)$:

$$Z(\mathbf{x}_0) = \mu + \sum_{i=1}^{k} \lambda_i [Z(\mathbf{x}_i) - \mu] \qquad (4.19)$$

where μ is the mean value of the parameter, k is the number of observations and λ_i are the weighting coefficients that satisfy the matrix equation

$$\begin{bmatrix} \mathrm{Cov}(\mathbf{x}_1,\mathbf{x}_1) & \mathrm{Cov}(\mathbf{x}_2,\mathbf{x}_1) & \ldots & \mathrm{Cov}(\mathbf{x}_k,\mathbf{x}_1) \\ \mathrm{Cov}(\mathbf{x}_1,\mathbf{x}_2) & \mathrm{Cov}(\mathbf{x}_2,\mathbf{x}_2) & \ldots & \mathrm{Cov}(\mathbf{x}_k,\mathbf{x}_2) \\ \ldots & \ldots & \ldots & \ldots \\ \mathrm{Cov}(\mathbf{x}_1,\mathbf{x}_k) & \mathrm{Cov}(\mathbf{x}_2,\mathbf{x}_k) & \ldots & \mathrm{Cov}(\mathbf{x}_k,\mathbf{x}_k) \end{bmatrix} \begin{bmatrix} \lambda_1 \\ \lambda_2 \\ \ldots \\ \lambda_k \end{bmatrix} = \begin{bmatrix} \mathrm{Cov}(\mathbf{x}_0,\mathbf{x}_1) \\ \mathrm{Cov}(\mathbf{x}_0,\mathbf{x}_2) \\ \ldots \\ \mathrm{Cov}(\mathbf{x}_0,\mathbf{x}_3) \end{bmatrix}$$
$$(4.20)$$

and the Cov operator denotes covariance, which is defined in terms of a semivariogram (discussed further in the next subsection).

The above method is often referred to as 'simple kriging', which requires knowing μ *a priori*. An alternative method is 'ordinary kriging' which seeks to improve on simple kriging by losing the need to estimate μ *a priori*. With ordinary kriging, $Z(x_0)$ is found from

$$Z(\mathbf{x}_0) = \sum_{i=1}^{k} \lambda_i Z(\mathbf{x}_i) \qquad (4.21)$$

where λ_i (and μ for that matter) now satisfy the matrix equation

$$\begin{bmatrix} \text{Cov}(\mathbf{x}_1,\mathbf{x}_1) & \text{Cov}(\mathbf{x}_2,\mathbf{x}_1) \dots \text{Cov}(\mathbf{x}_k,\mathbf{x}_1) & 1 \\ \text{Cov}(\mathbf{x}_1,\mathbf{x}_2) & \text{Cov}(\mathbf{x}_2,\mathbf{x}_2) \dots \text{Cov}(\mathbf{x}_k,\mathbf{x}_2) & 1 \\ \dots & \dots \quad \dots \quad \dots & \dots \\ \text{Cov}(\mathbf{x}_1,\mathbf{x}_k) & \text{Cov}(\mathbf{x}_2,\mathbf{x}_k) \dots \text{Cov}(\mathbf{x}_k,\mathbf{x}_k) & 1 \\ 1 & 1 \quad \dots \quad 1 & 0 \end{bmatrix} \begin{bmatrix} \lambda_1 \\ \lambda_2 \\ \dots \\ \lambda_k \\ \mu \end{bmatrix} = \begin{bmatrix} \text{Cov}(\mathbf{x}_0,\mathbf{x}_1) \\ \text{Cov}(\mathbf{x}_0,\mathbf{x}_2) \\ \dots \\ \text{Cov}(\mathbf{x}_0,\mathbf{x}_3) \\ 1 \end{bmatrix} \quad (4.22)$$

It can be appreciated that ordinary kriging takes longer to compute as it requires the solution of an additional equation at each point. In practice, the difference between ordinary and simple kriging is trivial. Another method is 'universal kriging', which improves on simple and ordinary kriging by relaxing the assumption that the mean is constant with space (see Olea, 1999, for more detail on this).

THE SEMIVARIOGRAM

The covariance $\text{Cov}(\mathbf{x}_i, \mathbf{x}_j)$ is found from

$$\text{Cov}(\mathbf{x}_i, \mathbf{x}_j) = \text{Cov}(0) - \gamma(h) \quad (4.23)$$

where h is the distance between points $\mathbf{x}_i$ and $\mathbf{x}_j$ and $\gamma(h)$ is the semivariogram defined by

$$\gamma(h) = \frac{1}{2}[Z(x) - Z(x+h)]^2 \quad (4.24)$$

The definition above involves the assumption that the semivariogram is not a smooth function of location and depends only on the separation of the pair of variates being considered. An important point is that the semivariogram is not a single number but a continuous function of the separation (sometimes referred as the 'lag'), h.

Generally, it is assumed that the semivariogram increases with lag over short lags and then ultimately levels off at some value $\text{Cov}(0)$. Therefore, it can be said that

$$\text{Cov}(0) = \lim_{h \to \infty} \gamma(h) \quad (4.25)$$

Direct application of equation (4.24) leads to the semivariogram cloud (a dense scatter of points). In practice, a sample variogram is used, found from

$$\gamma(h) = \frac{1}{2N_h} \sum_{\mathbf{x}_i - \mathbf{x}_j \approx h} [Z(\mathbf{x}_i) - Z(\mathbf{z}_j)]^2 \quad (4.26)$$

where N_h is the number of data pairs that satisfy $\mathbf{x}_i - \mathbf{x}_j \approx h$.

It can be easily appreciated that the sample semivariogram is a discontinuous function. Consequently, the sample semivariogram is approximated by fitting an analytical function. Popular functions include:

- the spherical semivariogram

$$\gamma(h) = \begin{cases} 0 & ,h = 0 \\ B + (C - B)\left[\dfrac{3h}{2a} - \dfrac{1}{2}\left(\dfrac{h}{a}\right)^3\right] & ,0 < h < a \\ C & ,h \geq a \end{cases} \quad (4.27)$$

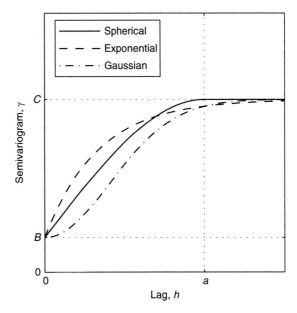

Figure 4.10. Variograms. C is the sill (total variance), B is the nugget (variance of the spatially uncorrelated portion of the field) and a is the range (distance over which the field is correlated).

- the exponential semivariogram

$$\gamma(h) = \begin{cases} 0 & ,h = 0 \\ B + (C - B)\left[1 - \exp\left(-\dfrac{3h}{a}\right)\right] & ,h > 0 \end{cases} \quad (4.28)$$

- the Gaussian semivariogram

$$\gamma(h) = \begin{cases} 0 & ,h = 0 \\ B + (C - B)\left[1 - \exp\left(-3\left(\dfrac{h}{a}\right)^2\right)\right] & ,h > 0 \end{cases} \quad (4.29)$$

where B is referred to as the nugget, C is the sill and a is the range.

The concept of the variogram sounds somewhat abstract but it is an excellent synthesis of the basic concepts one would like to bear in mind when dealing with spatial variability: scale and magnitude. As seen in Figure 4.10, the two basic properties of the variogram are the range and the sill. The range quantifies the distance over which properties are correlated (i.e. the spatial scale of heterogeneity). It can also be viewed as the average size of heterogeneous features. The sill is the variance of the field and, therefore, quantifies the magnitude of heterogeneity. There are cases in which one cannot define either scale (variability occurs at all scales) or magnitude (the variance increases with scale). In such cases one needs to use other types of variograms, which are conjectured to represent an ensemble of variability scales (Neuman, 1990).

CONDITIONAL SIMULATION

A problem with kriging is that, while it acknowledges uncertainty, it underestimates variability – that is, kriging estimates are too smooth. This also leads to the prediction of overly smooth features far away from the points of observation (see Figure 4.9c). Often, one is not as interested in obtaining the

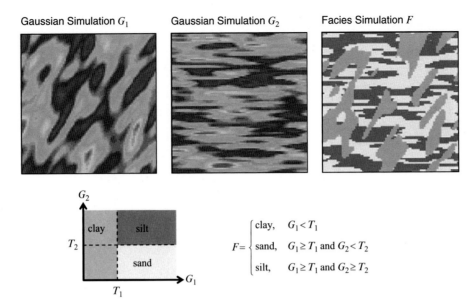

Figure 4.11. Gaussian threshold simulation method.

'optimum' estimate as in reproducing spatial variability. This is squarely the objective of conditional simulation. Here, 'simulation' refers to the generation of fields with the desired variability patterns, while 'conditional' refers to the constraint that any simulation should coincide with measured values at measurement points (see Figure 4.9d).

Sequential Gaussian simulation
Sequential Gaussian simulation (SGS) is one method that seeks to overcome these problems. The underlying principle is that the error variance at each estimated point, σ_{SK}, often referred to as the kriging variance, can be calculated from

$$\sigma_{SK}(\mathbf{x}_0) = \mathrm{Cov}(0) - \sum_{i=1}^{k} \lambda_i \mathrm{Cov}(\mathbf{x}_0, \mathbf{x}_i) \qquad (4.30)$$

With SGS, the value estimated by equation (4.19) or (4.21) is treated as the expectant. SGS then makes a stochastic estimate of the value by adding a residual sampled from a Gaussian distribution of variance, σ_{SK}^2, as obtained from equation (4.30) (hence Gaussian simulation). This value is then added to the collection of observations (i.e. the right-hand side of equation (4.19) or (4.21)). This procedure is repeated sequentially for each desired value of $\mathbf{x}_0$. It should be appreciated that the matrices in equation (4.20) or (4.22) also become increasingly larger with each estimate. Consequently, these calculations become very computationally intensive and one needs special algorithms to make them feasible (Gomez-Hernandez and Journel, 1993).

Facies simulation methods
A problem with traditional geostatistical simulation methods such as SGS is that they yield continuous fields. Geological reality is often viewed as discontinuous because different formations (facies) display different properties. Since formation boundaries are sharp, one should expect properties (i.e. the logarithm of hydraulic conductivity) to be discontinuous at interfaces. This is what motivates facies simulation methods.

The simplest and oldest facies simulation methods are the Boolean methods. These consist of simulating a number of objects (e.g. sand or shale lenses) of predefined shape and possibly random orientation, size and location. The result is indeed a number of geological bodies with the desired geometry and shape. The method was rather popular in the oil industry (Haldorsen and Chang, 1986) and in the fractured media hydrology literature (Long *et al.*, 1985). A drawback of Boolean methods is that they are not easy to condition on actual measurements. As such, they are appropriate for theoretical analyses of the effect of facies variability on groundwater flow and transport, but they are of limited use for simulating actual aquifers – hence the motivation for geostatistical facies simulation methods. The simplest one is the Gaussian threshold method.

The Gaussian threshold method (Galli *et al.*, 1994) is outlined in Figure 4.11. In essence it consists of the following steps:

1. Define a facies connection map that represents the allowed contacts between facies and the proportions of each. For example, in Figure 4.11, clay is allowed to contact sand and silt and occupies around one third of the volume.

2. Define the variograms of two variables (G_1 and G_2) such that they can represent the basic features of facies contacts. For example, the variogram of G_1 in Figure 4.11 is tilted because one wishes clay bodies to be tilted. On the other hand, the variogram of G_2 is much more anisotropic because sand and silt facies are elongated.

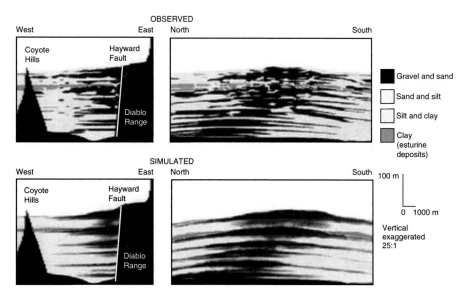

Figure 4.12. Cross-section of the San Francisco Bay as simulated by Koltermann and Gorelick (1996), using the genetic approach.

3. Perform a geostatistical simulation of G_1 and G_2 conditioned to known facies using, for example, the SGS method outlined above.
4. Map G_1 and G_2 onto the facies simulation using the facies connection map using a conditional relationship such as detailed in the bottom right of Figure 4.11.

The method is potentially powerful, but has not been widely used. Much more popular are the indicator simulation methods (Gomez-Hernandez and Journel, 1993). These are similar to SGS but work with 'indicator' variables, $I(\mathbf{x})$ (that is, $I(\mathbf{x}) = 1$ if $\mathbf{x}$ belongs to the simulated facies, and zero otherwise).

Another similar method is the Markov chain method (Carle and Fogg, 1996; Fogg *et al.*, 1998). It consist of defining transition probabilities (that is, $T_{i,j}(h)$ is the probability that $F(\mathbf{x} + h) = f_i$ if $F(\mathbf{x}) = f_i$) and using them to simulate the facies using a methodology similar to that for indicator simulations.

Genetic models
Facies simulation methods are effective in that they produce facies with the desired properties. However, the fields they produce are not based on geological reality – hence the motivation of genetic methods (Kolterman and Gorelick, 1992, 1996). Here, one attempts to simulate the processes that lead to the formation of geological bodies. This implies simulating past climate (i.e. floods), sediment transport processes (erosion, deposition), stream dynamics (meandering and such), tectonics (subsidence and tilting of sediments), etc. The effort in data acquisition, conceptual description and computer time is huge. The result, however, is quite impressive (Figure 4.12).

It is hard to imagine genetic models being used regularly in practice in the foreseeable future. However, this kind of model can be used to generate heterogeneity patterns, which can be incorporated into multiple point geostatistical methods (Strebelle, 2002; Liu and Journel, 2004) to generate facies and properties conditioned to actual data. Such methods are currently just emerging. It is hard to know how they will evolve, but they look hopeful in that one will eventually be capable of simulating properties that are both geologically reasonable and well conditioned on local data and understanding.

4.4.3 Upscaling

It should be apparent at this stage that variability is ubiquitous (all geological media display heterogeneity), that it is important (it affects flow and transport) and that it is hard to describe. In the previous sections, we have outlined a number of methods to deal with spatial variability. The immediate use of the fields generated with those methods would be to enter them as input to flow and transport simulators. We will do just that in the subsection 'Apparent and interpreted parameters' below. However, usually it is not practical because spatial variability occurs at many different scales. This implies that thorough descriptions of heterogeneity may require millions (if not billions) of nodes. Computational capabilities, however, restrict the number of nodes to a few thousand in most groundwater models. For each grid element, a single set of hydraulic parameters must be chosen that appropriately represent the subgrid spatial variability. This procedure is referred to as upscaling.

There is a vast array of different upscaling techniques within the literature. This large quantity of material is indicated by a correspondingly large number of review papers on the subject (Wen and Gomez-Hernandez, 1996; Renard and de Marsily, 1997; Wood, 2000; Farmer, 2002; Durlofsky, 2003; Noetinger *et al.*, 2005; Gueguen *et al.*, 2006; Sanchez-Vila *et al.*, 2006).

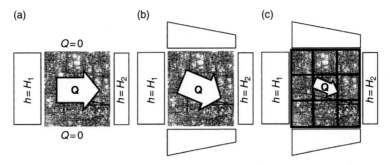

Figure 4.13. Alternative boundary conditions for computing equivalent transmissivity: (a) permeameter; (b) linearly varying; (c) border region.

Such a large number reflects the importance of the topic, but also that no technique is capable of fully solving the problem.

In this context, it is relevant to realise that upscaling may not be restricted to finding representative parameters as the scale (size of 'homogeneous' model block) increases, but may require changing the governing equations. In what follows, however, we will restrict our discussion to parameter upscaling.

EFFECTIVE PARAMETERS

Upscaling can be addressed by two broad families of methods, which we will term here 'continuous' and 'discrete'. Continuous methods start from the stochastic definition of the relevant variable (e.g. hydraulic conductivity) and use mathematical methods to find a theoretical solution. Actually, what they find is the 'effective' parameter. Effective parameters are those that represent the large-scale behaviour. For example, effective hydraulic conductivity is the ratio of mean (expected value) of flux to mean gradient. The earliest work on stochastic hydrogeology was aimed at seeking effective conductivity. Gutjahr *et al.* (1978) found that the relationship between effective and local permeability depends on the spatial dimension:

$$1D: \ K_{eff} = K_g(1 - \sigma_y^2/2) \tag{4.31}$$

$$2D: \ K_{eff} = K_g \tag{4.32}$$

$$3D: \ K_{eff} = K_g\left(1 + \sigma_y^2/6\right) \tag{4.33}$$

where K_g is the geometric average of local values and σ_y^2 is the variance of Y ($\ln K$).

Still more interesting is the case in which local conductivity is isotropic but its variogram is anisotropic. In this case, effective conductivity is given by (Gelhar and Axness, 1983; Dagan, 1989)

$$\mathbf{K}_{eff} = K_g\left[\mathbf{I} + \sigma_y^2\left(\frac{1}{2}\mathbf{I} - \mathbf{M}\right)\right] \tag{4.34}$$

where $\mathbf{M}$ depends on the assumed covariance function for $\ln K$. The most interesting feature of (4.34) is that, as shown by Dagan (1989) for the case of axisymmetric covariance, the term in parenthesis in (4.34) is positive-definite. This implies that directional effective permeability is larger than its local counterpart. In summary, the effective conductivity can display anisotropy even if the local conductivity is isotropic.

The scale dependence of dispersion has been one of the most spectacular results of stochastic hydrogeology. An effective dispersion coefficient has been found for large distances, even when no dispersion is assumed at the small scale. The effective dispersion – macrodispersion – coefficient is proportional to the variance of $\ln K$ and to its integral distance (Gelhar *et al.*, 1979; Gelhar and Axness, 1983; Dagan, 1982, 1984; Matheron and de Marsily, 1980; Neuman *et al.*, 1987; and many others).

EQUIVALENT PARAMETERS

The concept of effective parameter is both theoretically appealing and conceptually insightful. The above results explain why one of the most widely adopted strategies is to take the geometric average of transmissivity measurements (but keep reading!). However, the applicability of the concept to real situations is limited because it requires rather restricting assumptions of heterogeneity and Gaussianity that are rarely, if ever, met in practice. To deal with realistic situations it is much more convenient to work with equivalent parameters. In the case of transmissivity, it would be defined as follows. Given a real heterogeneous medium, its equivalent transmissivity is that of a homogeneous medium with the same geometry, which, subject to the same boundary conditions, allows the same amount of water to flow through.

This definition leads to the first and most basic approach to define equivalent transmissivity (Desbarats, 1987; Figure 4.13a). Assume that transmissivity is known (i.e. it has been simulated using any of the methods described in Section 4.4.2). Then for every block, repeat the following steps:

1. Isolate the block, and impose the boundary conditions of Figure 4.13a (prescribed heads on opposite boundaries and no flow on the other two).

2. Use any code to solve the flow equation and compute the flow rate across the domain.

3. Compute equivalent transmissivity, using Darcy's law, as

$$K_{eq} = (Q/L_y)[L_x/(H_1 - H_2)]$$

This approach is limited to rectangular blocks and can produce only anisotropy aligned with the axes. A somewhat more general alternative consists of prescribing linearly varying boundary conditions (Long et al., 1982; see Figure 4.13b). The method can now be applied to any boundary shape and rotating the mean head gradient yields tensorial transmissivities.

A common drawback with the two methods outlined above is that they ignore the variability surrounding the block. In fact, transmissivity is often said to be a 'non-local' property in the sense that its representative value depends on the connectivity with the surroundings. This problem is overcome by using a border region. In essence, the idea is to solve the flow equation (step 2 above) on not only the study block, but also in their surrounding blocks (see Figure 4.13c). Then the permeability of the block is defined as the ratio between volume averaged flux and gradient. That is (Rubin and Gomez-Hernandez, 1990):

$$\int_{block} q\,dv = K_{eq} \int_{block} \nabla h\,dv \qquad (4.35)$$

Other methods are available. These include the 'total dissipation energy' approach (Indelman and Dagan, 1993a, 1993b), where K_{eq} is defined as the hydraulic conductivity of a homogeneous medium that dissipates the same energy as the true heterogeneous medium. A simple yet effective method is the renormalisation approach of King (1989), which starts at the pixel level and associates pixels sequentially in series (adopting the harmonic average for the resulting pixel) and in parallel (arithmetic average). All these methods yield similar results (Sánchez-Vila et al., 1995), but no one is generally accepted as best.

An important result from this type of work is the finding that the equivalent transmissivity can be larger than the geometric average of point K values, whenever high K values are well connected (Sánchez-Vila et al., 1995). This is expected to be the rule, rather than the exception (think of fractured media, sedimentary materials with coarse sediments, palaeochannels, etc.). Therefore, one should expect K_{eq} to be larger than K_G. In fact, Knudby and Carrera (2005) use the ratio K_{eq}/K_G to define connectivity. The main question is whether one can estimate equivalent hydraulic conductivity from field data, which is addressed below.

APPARENT AND INTERPRETED PARAMETERS

In reality, one does not have at hand a pixel map of real aquifers. Therefore, the above methods would be of limited use if one cannot relate them to the type of data usually available in the field. Apparent and interpreted parameters are the ones obtained by interpreting field tests using conventional tools.

As such, these parameters are model dependent. The value of transmissivity one obtains from a pump test depends on whether the test is interpreted using the Theis, Jacob or Thiem method – that is, the resulting transmissivity is as much a property of the real medium as of the adopted model. The question is whether one can draw any general conclusion from such interpretation. This question was addressed by Meier et al. (1998). Their approach is summarised in Figure 4.14 and consists of the following steps:

1. Generate random fields. Both stationary Gaussian fields and well-connected fields were generated.
2. Simulate a pumping test by imposing a constant flow rate at one mode and observing drawdown at all other nodes.
3. Interpret drawdown curves at all nodes using Jacob's straight line method.

The main conclusions were as follows:

1. The resulting estimated transmissivities were identical at every node – that is, T estimated with Jacob's method is independent of the location of the observation point.
2. The estimated transmissivity was identical to the equivalent transmissivity (and, in the case of Gaussian fields, to the effective transmissivity).
3. The estimated storage coefficient reflects not only the actual storage coefficient, which was assumed constant, but also the degree of connection (high T) between the pumping well and the observation point.

4.5 CALIBRATION

Calibration is the process of evaluating model parameters such that model results fit the data as closely as possible. Calibration is rarely straightforward. Data come from various sources, with varying degrees of accuracy and levels of representativeness. Some parameters can be measured directly in the field, but such measurements are usually scarce, prone to error, and dependent on the measurement method used. Furthermore, since measurements are most often performed on scales and under conditions different from those required for modelling purposes, they tend to be both numerically and conceptually different from model parameters.

Calibration can be tedious and time-consuming because many combinations of parameters have to be evaluated, which also makes it prone to be incomplete. This, coupled with difficulties in taking into account the reliability of different pieces of information, makes it very hard to evaluate the quality of results. Therefore, it is not surprising that significant efforts have been devoted to the development of automatic calibration methods. Reviews of groundwater hydrology inversion are presented by

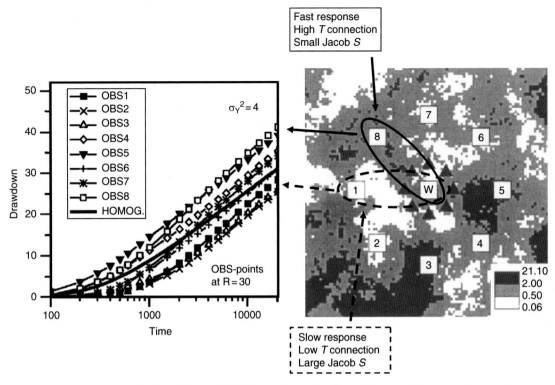

Figure 4.14. Summary of the approach taken by Meier *et al.* (1998) to evaluate the effect of heterogeneity on pumping tests interpretation. On the right-hand side is the 'true' transmissivity field showing the location of 8 observation points (numbered white squares) located at a distance of 30 m from a pumping well ('W' in a white square). On the left-hand side are the corresponding drawdown plots. They all display the same late time slope (i.e. yield the same transmissivity). However, response times (storativity) are highly variable.

Yeh (1986), Carrera (1988) and Carrera *et al.* (2005), and by Sorooshian and Gupta (1995) and Bastidas *et al.* (2002) for surface water.

4.5.1   Least squares and maximum likelihood

Traditionally, calibration techniques have largely been based on the method of maximum likelihood (see Ang and Tang, 1975). Let us declare the likelihood estimator of a trial model to be (McIntyre *et al.*, 2002)

$$L = P(\varepsilon_1)P(\varepsilon_2|\varepsilon_1)\prod_{i=3}^{N}P(\varepsilon_i|[\varepsilon_1 \cap \varepsilon_2 \cap \ldots \cap \varepsilon_{i-1}]) \qquad (4.36)$$

where ε_i is the ith of N model residuals (i.e. errors between modelled and observed data), $P(\varepsilon_i)$ is the probability density of ε_i and $P(\varepsilon_2|\varepsilon_1)$ is the probability of ε_2 assuming that ε_1 has already happened.

Definition of the likelihood function requires specifying the data vector (measurements of state variables, $\mathbf{h}$ and prior estimates of model parameters, $\bar{\mathbf{p}}$) and the error structure. If the residuals follow a zero-mean multivariate Gaussian distribution (which is an assumption), the error structure can be fully defined by the covariance matrices of the measurements and parameters,

$\mathbf{C}_{hh}$ and $\mathbf{C}_{pp}$ respectively. The likelihood function L then takes the form (Edwards, 1972)

$$L = \frac{\exp\left[-\frac{1}{2}(\mathbf{h}-\mathbf{h}(\mathbf{p}))^T\mathbf{C}_{hh}^{-1}(\mathbf{h}-\mathbf{h}(\mathbf{p})) - \frac{1}{2}(\mathbf{p}-\bar{\mathbf{p}})^T\mathbf{C}_{pp}^{-1}(\mathbf{p}-\bar{\mathbf{p}})\right]}{\sqrt{(2\pi)^{N_h+N_p}|\mathbf{C}_{hh}||\mathbf{C}_{pp}|}} \qquad (4.37)$$

where $\mathbf{h}(\mathbf{p})$ is the vector of corresponding computations (i.e. values of $\mathbf{h}$ computed with model parameters $\mathbf{p}$), and N_h and N_p are the number of data points and model parameters respectively. It follows that maximising the likelihood is equivalent to minimising the expression in the exponent

$$f \equiv (\mathbf{h}-\mathbf{h}(\mathbf{p}))^T\mathbf{C}_{hh}^{-1}(\mathbf{h}-\mathbf{h}(\mathbf{p})) + (\mathbf{p}-\bar{\mathbf{p}})^T\mathbf{C}_{pp}^{-1}(\mathbf{p}-\bar{\mathbf{p}}) \qquad (4.38)$$

Function f in equation (4.38) is often referred to as an objective function. Notice that f is a sum of squared error. Therefore minimising f is often termed 'finding the least squares'.

If the Gaussianity assumption is inappropriate, the statistical byproducts of the estimation should be used with caution. However, the method of obtaining optimal parameter sets by least squares itself is known to remain rather robust.

Differentiating with respect to $\mathbf{p}$ leads to

$$\frac{\partial f}{\partial \mathbf{p}} = -2\frac{\partial \mathbf{h}}{\partial \mathbf{p}}\mathbf{C}_{hh}^{-1}(\mathbf{h}-\mathbf{h}(\mathbf{p})) + 2\mathbf{C}_{pp}^{-1}(\mathbf{p}-\bar{\mathbf{p}}) \qquad (4.39)$$

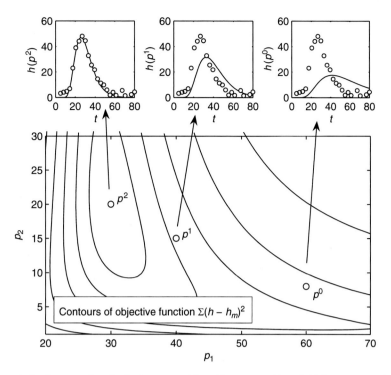

Figure 4.15. Schematic of the optimisation procedure in the inverse problem. One starts with an initial parameter set $\mathbf{p}^0$, which is used to compute heads $\mathbf{h}(\mathbf{p}^0)$.

which means the optimal parameter set, $\mathbf{p}_{\min}$, satisfies the conditional expectation formula

$$\mathbf{p}_{\min} = \overline{\mathbf{p}} + \mathbf{C}_{ph}\mathbf{C}_{hh}^{-1}(\mathbf{h} - \mathbf{h}(\mathbf{p}_{\min})) \qquad (4.40)$$

where $\mathbf{C}_{ph} \equiv \mathbf{C}_{pp}\mathbf{J}_{ph}$ is the cross-covariance between $\mathbf{h}$ and $\mathbf{p}$, and $\mathbf{J}_{ph} \equiv \partial\mathbf{h}/\partial\mathbf{p}$ is known as the Jacobian matrix. An approximation of the covariance matrix of estimated parameters can be obtained from

$$\sum\nolimits_{pp} = \left(\mathbf{C}_{pp}^{-1} + \mathbf{J}_{ph}^t \, \mathbf{C}_{hh}^{-1} \, \mathbf{J}_{ph}\right)^{-1} \qquad (4.41)$$

Carrera and Neuman (1986a, 1986b, 1986c) present further details on this methodology and discuss the advantages of using likelihood for model selection and network design.

OPTIMISATION ALGORITHMS
In practice, the maximum likelihood is obtained iteratively by introducing the linearised approximation

$$\mathbf{h}(\mathbf{p}^{K+1}) = \mathbf{h}(\mathbf{p}^K) + \mathbf{J}_{ph}(\mathbf{p}^{K+1} - \mathbf{p}^K) \qquad (4.42)$$

where K refers to the iteration number.

Substitution of equation (4.42) into (4.40) then leads to the explicit expression

$$\mathbf{p}^{K+1} = \mathbf{p}^K \left(\mathbf{C}_{pp}^{-1} + \mathbf{J}_{ph}\mathbf{C}_{hh}^{-1}\mathbf{J}_{ph}\right)^{-1} \left[\mathbf{J}_{ph}\mathbf{C}_{hh}^{-1}\left(\mathbf{h} - \mathbf{h}(\mathbf{p}^K)\right) + \mathbf{C}_{pp}^{-1}\left(\mathbf{p}^{K+1} - \mathbf{p}^K\right)\right] \quad (4.43)$$

which is the fundamental basis for the commonly used optimisation software packages PEST and UCODE.

The algorithm then proceeds as illustrated in Figure 4.15. One starts at an initial point $\mathbf{p}^0$, computes model outputs at measurement points and evaluates the objective function (4.38). Then, equation (4.43) is used to obtain an improved estimate of $\mathbf{p}^1$, and the procedure is repeated until convergence is achieved.

4.5.2 Global optimisation

It can be easily appreciated that the above method is profoundly dependent on the choice of $\mathbf{p}^0$. Techniques such as those described above are often referred to as local optimisation routines because they are capable only of finding the stationary points of f in close proximity to $\mathbf{p}^0$. In situations when the response surface $f(\mathbf{p})$ is very noisy it is arguably better to utilise a selection of $\mathbf{p}^0$ estimates obtained by randomly sampling an a-priori distribution or by trial and error (Carrera and Neuman, 1986a, 1986b, 1986c). This method is sometimes referred to as the multi-start local optimisation method (Duan, 2003).

Another method is the genetic algorithm (Holland, 1975). This procedure can be described as follows:

1. Randomly generate a population of parameter sets in the feasible parameter space.

2. Rank the population according to corresponding objective function values.

3. Choose two sets from the population to make two offspring using genetic operators (e.g. cross-over and mutate).

4. Exchange the two worst sets with the two offspring.

5. Continue steps 3 and 4 until the statistical distribution of the population has converged.

An increasingly popular method in hydrology is the shuffled complex evolution (SCE) algorithm (Duan *et al.*, 1993) and variations (Yapo *et al.*, 1998; Vrugt *et al.*, 2003, 2005), which combine the benefits of local search methods and evolution algorithms. The basic procedure is as follows:

1. Randomly generate a population of parameter sets in the feasible parameter space.

2. Rank the population according to corresponding objective function values.

3. Partition the population into a number of complexes.

4. Generate an evolved parameter set from each using the simplex method.

5. Replace the worst parameter set in each complex with the corresponding evolved one.

6. Reconcile complexes and rank according to objective function values.

7. Continue steps 3 to 6 until the statistical distribution of the population has converged.

Another interesting method is simulated annealing (SA) (e.g. Thyer *et al.*, 1999; Sumner *et al.*, 1997). In physical annealing a metal is heated to a temperature just below its melting point and then cooled slowly enough to allow the molecules to align in a minimum energy state. The SA calibration procedure involves randomly searching around continuously improving parameter estimates. As the number of iterations increases, the distance of the random searches is reduced (as with the temperature in physical annealing) until ultimately the system converges. The slower the cooling rate, the higher the chance of finding the 'global optimum' but the longer it takes to complete.

Of course, we can never be sure that a global optimum exists. It is often the case that many different representations of a hydrological system are equally valid in terms of their ability to produce acceptable simulations of available data (Beven, 2006).

4.5.3 Sensitivity analysis

Sensitivity analysis refers to the assessment of the dependence between model inputs and outputs and, specifically, how the latter respond to perturbations in the former. In mathematical terms, the sensitivity of a model output h to a model parameter p is usually defined by the derivative of h in respect to p. This will be contained within the Jacobian matrix $\mathbf{J}_{ph}$ mentioned above. If the derivative is very large, then small perturbations in p will lead to large perturbations in h. Conversely, a small derivative implies that h is relatively less sensitive to p. High sensitivities relative to the knowledge of parameters $\mathbf{p}$ imply high uncertainties.

For large, transient, spatially distributed groundwater models, it is often not feasible to compute the derivative of every state variable at every cell and every time with respect to every input variable. Therefore, care must be taken in deciding which effects and factors to consider within the sensitivity analysis. To start with, one should consider the main objectives of the modelling exercise. The process of analysing model output and sensitivities may then shed light on additional important outputs. For example, if model output suggests that groundwater flooding is important one should additionally study maximum water levels.

In practice (PEST, UCODE, etc.), the Jacobian matrix is approximated by finite differences (recall Section 4.2.3) whereby

$$\frac{\partial h}{\partial p_i} \approx \frac{h(p_i + \Delta p_i) - h(p_i)}{\Delta p_i}, \; i = 1 \dots N_p \qquad (4.44)$$

Unfortunately, in many cases the derivative of $\mathbf{h}$ with respect to $\mathbf{p}$ is discontinuous. Even when it is continuous, non-linearity may be so marked that the concept of sensitivity as a derivative is of limited validity. Therefore, it is often convenient to redefine sensitivity as the variation in model output in response to changes in model parameters within their range of reasonable values.

A sensitivity analysis can provide qualitative preliminary information concerning the uncertainty of model calibrations and subsequent predictions. However, this is no substitute for a rigorous uncertainty analysis.

4.6   UNCERTAINTY PROPAGATION

An important but often ignored issue in groundwater modelling is the analysis of uncertainty associated with predicted effects (model outputs $\mathbf{f}$). Broadly speaking, uncertainty analysis can be defined as 'the means of calculating and representing the certainty with which the model results represent reality' (McIntyre *et al.*, 2002). This uncertainty is caused by the fact that model inputs, $\mathbf{p}$ are uncertain. As can be appreciated from the material above, sources of uncertainty are plentiful including (McIntyre *et al.*, 2002):

- model parameter error;
- model structure error (where the model structure is the set of numerical equations which define the uncalibrated model);
- numerical errors, truncation errors, rounding errors and typographical mistakes in the numerical implementation;
- boundary condition uncertainties;
- sampling errors (i.e. the data not representing the required spatial and temporal averages);
- measurement errors (e.g. due to methods of handling and laboratory analysis);
- human reliability.

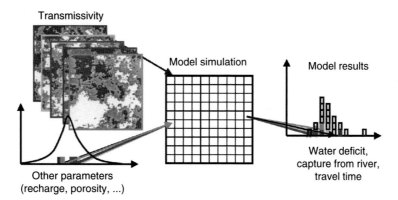

Figure 4.16. In the Monte Carlo method, uncertainty in model outputs is evaluated by repeated simulations with varying inputs according to their uncertainty.

If **f** is a model prediction to be made using input data **p**, then a lower bound estimate of its variance due to parameter uncertainty can be obtained from

$$\mathrm{Var}(\mathbf{f}) = \left(\frac{\partial \mathbf{f}}{\partial \mathbf{p}}\right)^{\mathrm{T}} \Sigma_p \left(\frac{\partial \mathbf{f}}{\partial \mathbf{p}}\right) \tag{4.45}$$

where Σ_p is yet another covariance matrix.

Although the real uncertainty is likely to be much larger, the above equation allows a quantitative exploration of how different factors affect prediction uncertainty. However, similar information is gained from a rigorous sensitivity analysis. To properly explore prediction uncertainty, more sophisticated techniques are needed.

Liu and Gupta (2007) list a variety of uncertainty analysis frameworks that have been introduced in the hydrologic literature including:

- the generalised likelihood uncertainty estimation (GLUE) methodology (Beven and Binley, 1992);
- the Bayesian recursive estimation technique (BaRE) (Thiemann *et al.*, 2001);
- the maximum likelihood Bayesian averaging method (MLBMA) (Neuman, 2003);
- the simultaneous optimisation and data assimilation algorithm (SODA) (Vrugt *et al.*, 2005).

At the heart of these are variations of the Monte Carlo method (MCM). In essence, MCM consists of evaluating **h** repeatedly for a large number of combinations of model inputs **p**. This process can be described as follows (Figure 4.16):

1. Randomly sample a vector of model inputs, $\mathbf{p}_n$, from pre-defined probability distributions.
2. Use the model to compute the desired output, $\mathbf{f}(\mathbf{p}_n)$.
3. Add model output $\mathbf{f}(\mathbf{p}_n)$ to existing population, $\mathbf{f}(\mathbf{p}_{i=1.n-1})$.
4. Continue steps 1 to 3 until $\mathbf{f}(\mathbf{p}_i)$ statistically converges.

A major issue is in determining which statistical descriptors are relevant to the task at hand. For example, Figure 4.17 shows 200

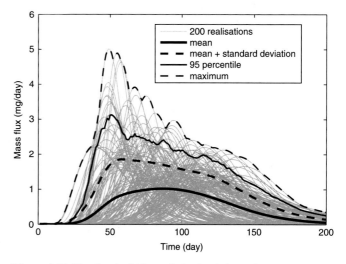

Figure 4.17. Two hundred Monte Carlo simulations of a breakthrough curve. Notice that the ensemble average (mean of all realisations) severely underestimates peak concentrations.

Monte Carlo simulations of a solute breakthrough curve. Both the velocity and diffusion coefficients were treated as independent random variables. A commonly used statistic for a wide variety of applications is the mean. However, in Figure 4.17 it can be seen that while individual realisations convey an idea of uncertainty in travel time and maximum concentration, their ensemble average does not. One should consider alternative statistics (e.g. standard deviation, 95 percentile and maximum value).

It may also be necessary to examine statistical descriptions of specific features (peak value, time to peak, first arrival time etc.). Unfortunately this is not always clear *a priori*. Therefore, it is imperative to think ahead as to what might be expected. This can be aided by thoroughly examining the model output of a selected sample of simulations, in particular those simulations deemed to be extreme (e.g. those with highest peak or earliest arrival in Figure 4.17). Such a process also allows the modeller to gain further insight in to how their model behaves.

4.7 SUMMARY

The fundamental techniques for developing models for groundwater flow have been presented and reviewed. The chapter commenced with a discussion concerning the purpose of models. In this way, a distinction was made between site-specific models and generic models. Ideally a site-specific model should be for an aquifer management scheme what a financial accounting scheme is for a well-managed company. However, because of the difficulties associated with building and maintaining models, coupled with the legal and practical difficulties of managing aquifers in real time, this is rarely the case. Instead, models are treated as approximate and used as decision support tools. Although such models should not be expected to produce precise forecasts, they can provide reasonable assessments of different management scenarios. In this way, the relative advantages and disadvantages of alternative scenarios can be evaluated and the options ranked in a transparent and objective manner.

The chapter went on to discuss the relevant governing equations for groundwater flow modelling. Procedures for numerical solution were then presented, with a special emphasis on the use of finite differences. Following from this, a review of the various phases associated with modelling procedure was presented. These included conceptualisation, discretisation, calibration, model selection and uncertainty estimation.

The key parameters associated with groundwater flow, permeability, porosity, soil texture, etc. are highly variable in space. Variability is important not only because of the uncertainty it brings about, but also because large-scale behaviour of a spatially variable phenomenon may be significantly different from the small-scale behaviour. Consequently, the chapter dedicated a large section to the handling of spatial variability. Here, many aspects of geostatistics were introduced, including the semivariogram, kriging and sequential Gaussian simulation. This led to a discussion on the necessity of upscaling. For each model grid element, a single set of hydraulic parameters must be chosen that appropriately represent the subgrid spatial variability. In this context, the important terms of effective, equivalent and apparent parameters were compared and contrasted.

Following from this, the chapter focused on the topic of calibration. Calibration is the process of evaluating model parameters such that model results fit observed data as closely as possible. The discussion started with a derivation of the maximum likelihood method approach. A selection of global optimisation schemes were then presented, including the genetic algorithm, the shuffled complex evolution algorithm and simulated annealing. The section concluded with the important reminder that different representations of a hydrological system are often equally valid in terms of their ability to produce acceptable simulations of available data.

Finally, the important topic of uncertainty analysis was addressed. Starting with a simple listing of the many factors that give rise to model prediction uncertainty, the section then went on to describe how Monte Carlo methods can be used to develop quantitative frameworks for uncertainty estimation.

We have argued that groundwater modelling is an essential tool to integrate the information needed for groundwater management. However, over the last two decades, the associated literature has become voluminous, and it is easy for the user to be overwhelmed by the complexity of new techniques, and to lose sight of the essentials of the modelling process: the development, numerical implementation and testing of an appropriate conceptual model to guide the management process. Our aim here has been to provide an overview that puts recent developments into context, to point out the strengths and weaknesses of alternative techniques, and to point the reader to the source literature, as required. Hopefully, we have succeeded in providing some useful insights into the modelling process and some practical advice concerning the utility of these latest developments for future water management projects and scenarios.

REFERENCES

Anderson, M. P. and Woessner, W. W. (1992) *Applied Groundwater Modeling*. Academic Press.

Ang, A. H. S. and Tang, W. H. (1975) *Probability Concepts in Engineering Planning and Design, Vol. 1, Basic Principles*. Wiley.

Ayora, C., Tabener, C., Saaltink, M. W. and Carrera, J. (1998) The genesis of dedolomites: a discussion based on reactive transport modeling. *J. Hydrol.* **209**, 346–365.

Bastidas, L. A., Gupta, H. V. and Sorooshian, S. (2002) Emerging paradigms in the calibration of hydrologic models. In *Mathematical Models of Large Watershed Hydrology*, ed. V. P. Singh and D. K. Frevert, 23–87. Water Resources Publications.

Bear, J. (1979) *Hydraulics of Groundwater*. McGraw-Hill.

Beven, K. J. (2006) A manifesto for the equifinality thesis. *J. Hydrol.* **320**, 18–36.

Beven, K. J. and Binley, A. M. (1992) The future of distributed models: model calibration and uncertainty prediction. *Hydrological Processes* **6**, 279–298.

Carle, S. F. and Fogg, G. E. (1996) Transition probability-based indicator geostatistics. *Math. Geol.* **28**(4), 453–476.

Carrera, J. (1988) State of the art of the inverse problem applied to the flow and solute transport equations. In *Groundwater Flow and Quality Modelling*, 549–583. D. Reidel.

Carrera, J. (1993) An overview of uncertainties in modeling groundwater solute transport. *J. Contam. Hydrol.* **13**, 23–48.

Carrera, J. and Neuman, S. P. (1986a) Estimation of aquifer parameters under transient and steady-state conditions. 1. Maximum likelihood method incorporating prior information. *Water. Resour. Res.* **22**(2), 199–210.

Carrera, J. and Neuman, S. P. (1986b) Estimation of aquifer parameters under transient and steady-state conditions. 2. Uniqueness, stability and solution algorithms. *Water. Resour. Res.* **22**(2), 211–227.

Carrera, J. and Neuman, S. P. (1986c) Estimation of aquifer parameters under transient and steady-state conditions. 3. Application to Synthetic and Field Data. *Water. Resour. Res.* **22**(2), 228–242.

Carrera, J., Mousavi, S. F., Usunoff, E., Sanchez-Vila, X. and Galarza, G. (1993) A discussion on validation of hydrogeological models. *Reliability Engineering and System Safety* **42**, 201–216.

Carrera, J., Sànchez-Vila, X., Benet, I. *et al.* (1998) On matrix diffusion: formulations, solution methods and qualitative effects. *Hydrogeol. J.* **6**, 178–190.

Carrera, J., Alcolea, A., Medina, A., Hidalgo, J. and Slooten, L. J. (2005) Inverse problem in hydrogeology. *Hydrogeol. J.* **13**, 206–222.

Castro, A., Vazquez-Sune, E., Carrera, J., Jaen, M. and Salvany, J. M. (1999) Calibracion del modelo regional de flujo subterraneo en la zona de Aznalcollar, Espana: ajuste de las extracciones [Calibration of the groundwater flow regional model in the Aznalcollar site, Spain: extractions fit]. In *Hidrologa Subterranea. II, 13*. ed A. Tineo. Congreso Argentino de Hidrogeologia y IV Seminario Hispano Argentino sobre temas actuales de la hidrogeologia.

Celia, M. A., Russell, T. F., Herrera, I. and Ewing, R. E. (1990) An Eulerian–Lagrangian localized adjoin method for the advection–diffusion equation. *Adv. Water Resour.* **13**, 187–206.

Dagan, G. (1982) Stochastic modeling of groundwater flow by unconditional and conditional probabilities, 1. Conditional simulation and the direct problem. *Water Resources. Res.* **18**(4), 813–833.

Dagan, G. (1984) Solute transport in heterogeneous porous formations. *J. Fluid Mech.* **145**, 151–177.

Dagan, G. (1989) *Flow and Transport in Porous Formations*. Springer, New York.

de Marsily, G., Delay, F., Gonçalvès, J. *et al.* (2005) Dealing with spatial heterogeneity, *Hydrogeol. J.* **13**(1), 161–183.

Desbarats, A. J. (1987) Numerical estimation of effective permeability in sand-shale formation. *Water Resour. Res.* **23**(2), 273–286.

Diersch, H. J. G. (2005) *FEFLOW Reference Manual, WASY GmbH*, Institute for Water Resources Planning and Systems Research, Berlin.

Duan, Q., Gupta, V. K. and Sorooshian, S. (1993) A shuffled complex evolution approach for effective and efficient global minimization. *J. Optim. Theory Appl.* **76**(3), 501–521.

Duan, Q. (2003) Global optimization for watershed model calibration. In *Calibration of Watershed Models*, ed. Q. Duan, H. V. Gupta, S. Sorooshian, A. N. Rousseau and R. Turcotte. Water Science and Application **6**, 89–104. Am. Geophys. Union.

Durlofsky, L. J. (2003) Upscaling of geocellular models for reservoir flow simulation: a review of recent progress. *Proceedings of the 7th International Forum on Reservoir Simulation*, Bühl/Baden-Baden, Germany, 23–27 June 2003.

Edwards, A. (1972) *Likelihood*. Cambridge Univ. Press.

Farmer, C. L. (2002) Upscaling: a review. *Int. J. Numer. Meth. Fluids.* **40**, 63–78.

Fogg, G. E., Noyes, C. D. and Carle, S. F. (1998) Geologically based model of heterogeneous hydraulic conductivity in an alluvial setting. *Hydrogeol. J.* **6**, 131–143.

Freeze, R. A. and Cherry, J. A. (1979) *Groundwater*. Prentice-Hall.

Freeze, R. A. and Witherspoon, P. A. (1966) Theoretical analysis of regional groundwater flow. II. Effect of water table configuration and subsurface permeability variations. *Water Resour. Res.* **3**, 623–634.

Galli A., Beucher, H., Le Loch, G. and Doligez, B. (1994) The pros and cons of the truncated Gaussian method. In *Geostatistical Simulations*, ed. M. Armstrong and P. A. Dowd, 217–233. Kluwer.

Garven, G., and Freeze, R. A. (1984) Theoretical analysis of the role of groundwater flow in the genesis of stratabound ore deposits. II. Quantitative results. *Amer. J. Sci.* **284**, 1125–1174.

Gelhar, L. W., Gutjahr, A. L. and Naff, R. L. (1979) Stochastic analysis of macrodispersion in a stratified aquifer. *Water Resour. Res.* **15**, 1387–1397.

Gelhar, L. W. and Axness, C. L. (1983) Three Dimensional Stochastic Analysis of Macrodispersion in Aquifers. *Water Resour. Res.* **19**, 161–180.

Gelhar, L. W. (1986) Stochastic subsurface hydrology from theory to applications. *Water Resour. Res.* **22**(9), 135s–145s.

Gomez-Hernandez, J. J. and Journel, A. (1993) Joint sequential simulation of multigaussian fields. *Geostat Troia.* **1**, 85–94.

Gorelick, S. (1983) A review of distributed parameter groundwater management modeling methods. *Water Resour. Res.* **19**, 305–319.

Gueguen, Y., Le Ravalec, M. and Ricard, L. (2006) Upscaling: effective medium theory, numerical methods and the fractal dream. *Pure Appl. Geophys.* **163**, 1175–1192.

Gutjahr, A., Gelhar, L., Bakr, A. and MacMillan, J. (1978) Stochastic analysis of spatial variability in subsurface flows. 2. Evaluation and application, *Water Resour. Res.* **14**(5), 953–959.

Haldorsen, H. H. and Chang, D. M. (1986) Notes on stochastic shales from outcrop to simulation models. In *Reservoir characterization*. ed. L. W. Lake and H. B. Carol Jr., 152–167. Academic.

Heinrich, J. C., Huyakorn, P. S., Mitchell, A. R. and Zienkiewicz, O. C. (1977) An upwind finite element scheme for two-dimensional transport equation. *Int. J. Num. Meth. Eng.* **11**, 131–143.

Hill, M. C. (1992) A computer program (MODFLOWP) for estimating parameters of a transient, three dimensional, groundwater flow model using nonlinear regression. *USGS open-file report* 91–484.

Holland, J. H. (1975) Genetic Algorithms, computer programs that 'evolve' in way that resemble even their creators do not fully understand. *Scientific American*, July 1975, 66–72.

Indelman, P. and Dagan, G. (1993a) Upscaling of permeability of anisotropic heterogeneous formations. 1. The general framework. *Water Resour. Res.* **29**(4), 917–923.

Indelman, P. and Dagan, G. (1993b) Upscaling of permeability of anisotropic heterogeneous formations. 2. General structure and small perturbation analysis. *Water Resour. Res.* **29**(4), 925–933.

King, P. R. (1989) The use of renormalization for calculating effective permeability. *Transport in Porous Media* **4**, 37–58.

Kipp, K. L. (1987) HST3D: a computer code for simulation of heat and solute transport in three dimensional groundwater flow systems. *USGS Water Recourses Inv. Rep.* 86–4095.

Knudby, C. and Carrera, J. (2005) On the relationship between indicators of geostatistical, flow and transport connectivity. *Adv. Water. Resour.* **28**, 405–421.

Kolterman, C. E. and Gorelick, S. M. (1992) Paleoclimatic signature in terrestrial flood deposits. *Science* **256**, 1775–1782.

Kolterman, C. E. and Gorelick, S. M. (1996) Heterogeneity in sedimentary deposits: a review of structure-imitating, process-imitating, and descriptive approaches. *Water Resour. Res.* **32**, 2617–2658.

Konikow, L. F. and Bredehoeft, J. D. (1978) Computer model of two-dimensional solute transport and dispersion in ground water. *Techniques of Water Resources Investigations of the USGS*, Book 7, Chapter C2.

Liu, Y. and Gupta, H. V. (2007) Uncertainty in hydrologic modeling: Toward an integrated data assimilation framework. *Water Resour. Res.* **43**, W07401.

Liu, Y. H, and Journel, A. (2004) Improving sequential simulation with a structured path guided by information content. *Math. Geol.* **36**(8), 945–964.

Loaiciga, H. and Marino, M. (1987) The inverse problem for confined aquifer flow: identification and estimation with extensions. *Water Resour. Res.* **23**(1), 92–104.

Long, J., Remer, J., Wilson, C. and Witherspoon, P. (1982) Porous media equivalents for networks of discontinuous fractures. *Water Resour. Res.* **18**(3), 645–658.

Long, J., Gilmour, P. and Witherspoon, P. (1985) A model for steady fluid flow in random three-dimensional networks of disc-shaped fractures. *Water Resour. Res.* **21**(8), 1105–1115.

Martinez-Landa, L. and Carrera, J. (2005) An analysis of hydraulic conductivity scale effects in granite (Full-scale Engineered Barrier Experiment (FEBEX), Grimsel, Switzerland). *Water Resour. Res.* **41**, W03006, doi:10.1029/2004WR003458.

Mathias, S. A., Butler, A. P. and Zhan, H. (2008) Approximate solutions for Forchheimer flow to a well. *J. Hydraul. Eng.* **134**(9), 1318–1325.

Matheron, G. (1971) *The Theory of Regionalized Variables and Its Applications*. Les Cahiers du CMM, Fasc. No 5, ENSMP, Paris.

Matheron, G. and de Marsily, G. (1980) Is transport in porous media always diffusive? A counterexample. *Water Resour. Res.* **16**, 901–917.

McDonald, M. G. and Harbaugh, A. W. (1988) A modular three-dimensional finite difference groundwater flow model. *Techniques of Water Resources Investigations of the USGS*, Book 6, Chapter A1.

McIntyre, N., Wheater, H. S. and Lees, M. J. (2002) Estimation and propagation of parametric uncertainty in environmental models. *J. Hydroinform.* **4**(3), 177–198.

Meier, P., Carrera, J. and Sanchez-Vila, X. (1998) An evaluation of Jacob's method work for the interpretation of pumping tests in heterogeneous formations. *Water Resour. Res.* **34**(5), 1011–1025.

NEA-SKI (1990) *The International HYDROCOIN Project, Level 2: Model Validation*. OECD.

Neuman, S. P. (1981) A Eulerian-Lagrangian numerical scheme for the dispersion-convection equation using conjugate space-time grids. *J. Comput. Phys.* **41**, 270–294.

Neuman, S. P. (1990) Universal scaling of hydraulic conductivities and dispersivities in geologic media. *Water Resour. Res.* **26**(8), 1749–1758.

Neuman, S. P. (2003) Maximum likelihood Bayesian averaging of uncertain model predictions. *Stochastic Environ. Res. Risk Assess.* **17**, 291–305.

Neuman S. P., Winter, C. L. and Newman, C. M. (1987) Stochastic theory of field-scale fickian dispersion in anisotropic porous-media. *Water Resour. Res.* **23**(3), 453–466.

Noetinger, B., Artus V. and Zargar, G. (2005) The future of stochastic and upscaling methods in hydrogeology. *Hydrogeol J.* **13**(1), 184–201.

Olea, R. A. (1999) *Geostatistics for Engineers and Earth Scientists.* Kluwer Academic.

Pruess, K., Oldenburg, C. M. and Moridis, G. (1999) *TOUGH2 user's guide, version 2.0. Report LBNL-43134*, Lawrence Berkeley National Laboratory.

Renard, Ph. and de Marsily, G. (1997) Calculating equivalent permeability: a review. *Adv. Water Resour.* **20**(5–6), 253–278.

Richtmyer, R. D. and Morton, K. W. (1957) *Difference Methods for Initial Value Problems.* Interscience.

Rubin, Y. and Gomez-Hernandez, J. J. (1990) A stochastic approach to the problem of upscaling of conductivity in disordered media: theory and unconditional numerical simulations. *Water Resour. Res.* **26**(4), 691–701.

Sanchez-Vila, X., Girardi, J. and Carrera, J. (1995) A synthesis of approaches to upscaling of hydraulic conductivities. *Water Resour. Res.* **31**(4), 867–882.

Sanchez-Vila, X., Guadagnini, A. and Carrera, J. (2006) Representative hydraulic conductivities in saturated groundwater flow. *Rev. Geophys.* **44**, RG3002.

Stallman, R. W. (1956) Numerical analysis of regional water levels to define aquifer hydrology. *Transactions American Geophysical Union.* **37**(4), 451–460.

Sorooshian, S. and Gupta, V. K. (1995) Model calibration. In *Computer Models of Watershed Hydrology*, ed. V. P. Singh, 23–68. Water Resources Publications.

Steefel, C. I., DePaolo, D. J. and Lichtner, P. C. (2005) Reactive transport modeling: An essential tool and a new research approach for the Earth sciences. *Earth and Planetary Science Letters* **240**(3–4), 539–558.

Strebelle, S. (2002) Conditional simulation of complex geological structures using multiple point statistics. *Math. Geol.* **34**(1), 1–22.

Sumner, N. R., Flemming, P. M. and Bates, B. C. (1997) Calibration of a modified SFB model for twenty-five Australian catchments using simulated annealing. *J. Hydrol.* **197**, 166–188.

Thiemann, T., Trosset, M., Gupta, H. and Sorooshian, S. (2001) Bayesian recursive parameter estimation for hydrologic models. *Water Resour. Res.* **37**(10), 2521–2535.

Thyer, M., Kuczera, G. and Bates, B. C. (1999) Probabilistic optimization for conceptual rainfall–runoff models: a comparison of shuffled complex evolution and simulated annealing algorithms. *Water Resour. Res.* **35**(3), 767–773.

Vrugt, J. A., Gupta, H. V., Bouten, W. and Sorooshian, S. (2003) A Shuffled Complex Evolution Metropolis algorithm for optimization and uncertainty assessment of hydrologic model parameters. *Water Resour. Res.* **39**(8), 1201.

Vrugt, J. A., Diks, C. G. H., Gupta, H. V., Bouten, W. and Verstraten, J. M. (2005) Improved treatment of uncertainty in hydrologic modeling: combining the strengths of global optimization and data assimilation. *Water Resour. Res.* **41**, W01017.

Wen, X.-H. and Gomez-Hernandez, J. J. (1996) Upscaling hydraulic conductivities in heterogeneous media: an overview. *J. Hydrol.* **183**, ix–4ii.

Wood, B. D. (2000) *Review of Upscaling Methods for Describing Unsaturated Flow.* Pacific Northwest National Lab., Richland, WA (US) PNNL-13325.

Wu, Y. S. (2002) Numerical simulation of single-phase and multiphase non-Darcy flow in porous and fractured reservoirs. *Transp. Porous Media* **49**(2), 1573–1634.

Yapo, P. O., Gupta, H. V. and Sorooshian, S. (1998) Multi-objective global optimization for hydrologic models. *J. Hydrol.* **204**, 83–97.

Yeh, W. W.-G. (1986) Review of parameter identification procedures in groundwater hydrology: the inverse problem. *Water Resour. Res.* **22**(2), 95–108.

5 Performing unbiased groundwater modelling: application of the theory of regionalised variables

S. Ahmed, S. Sarah, A. Nabi and S. Owais

5.1 INTRODUCTION

Groundwater in arid and semi-arid regions is increasingly over-exploited because of high population growth and extensive agricultural application. It is essential to have a thorough understanding of the complex processes involved when undertaking groundwater assessment and management activities. Groundwater models play an important role due to their ability to estimate head distributions and flow rates for possible future scenarios. These models are computer-based numerical solutions to the boundary value problems of concern.

There is no doubt that the mathematical and computational aspect of groundwater modelling has reached a satisfactory level of development. Therefore, the focus of research in recent years has shifted to the problems of parameter identification and uncertainty quantification. Hydrogeological parameters display a large spatial heterogeneity, with possible variations of several orders of magnitude within a short distance. This spatial variability is difficult to characterise in a deterministic way. However, a statistical analysis shows that hydrogeological parameters do not vary in space in a purely random fashion. There is some structure to this spatial variability that can be characterised in a statistical way.

During the last few decades, numerous mathematical approaches have been used to estimate hydrogeological parameters from scattered or sparse data. The study of regionalised variables starts from the ability to interpolate a given field from only a limited number of observations while preserving the theoretical spatial correlation. This is accomplished by means of a geostatistical technique called kriging. The advantage of using geostatistics as an estimation technique is that the variance of the estimation error can be calculated at any point (where there may or may not already be an observed value) based on the geometry of the measurement network. Compared with other interpolation techniques, kriging is advantageous because it considers the number and spatial configuration of the observation points, the position of the data point within the region of interest, the distances between the data points with respect to the area of interest and the spatial continuity of the interpolated variable.

This chapter describes the theory of regionalised variables and how it can be applied to groundwater modelling studies, followed by a case study from the Maheshwaram watershed of Andhra Pradesh, India.

5.2 THE THEORY OF REGIONALISED VARIABLES: GEOSTATISTICS

Geostatistics was briefly covered in the previous chapter. However, it is useful to reiterate some key aspects. Geostatistics is a statistical technique based on the theory of regionalised variables. It is used to describe spatial relationships existing within or among several data sets. Developed by Matheron (1965), it was initially intended for the mining industry as high costs of drilling made the analysis of the data extremely important. An important problem faced in mining is the prediction of ore grade within a mining block from observed samples that are irregularly spaced. A fundamental tool in geostatistics is the variogram, which is used to quantify spatial correlations between observations. The estimation procedure kriging is named after D. Krige, who with his colleagues applied a range of statistical techniques to the estimation of ore reserves in the 1950s. Due to the many advantages of kriging, its application can be found in very different disciplines ranging from mining and geology to soil science, hydrology, meteorology, environmental sciences, agriculture, etc. Kriging can be applied wherever a measure is made on a sample at a particular location in space and/or time (Cressie, 1991; Armstrong, 1998). Kriging is best suited for spatially dependent variables, where it is necessary to use the limited data in order to infer the real nature of their spatial dependence as precisely as possible.

Groundwater Modelling in Arid and Semi-Arid Areas, ed. Howard S. Wheater, Simon A. Mathias and Xin Li. Published by Cambridge University Press.
© Cambridge University Press 2010.

The application of kriging to groundwater hydrology was initiated by Delhomme (1976, 1978, 1979) followed by a multitude of other researchers (e.g. Delfiner and Delhomme, 1983; de Marsily *et al.*, 1984; de Marsily, 1986; Aboufirassi and Marino, 1983, 1984; Gambolti and Volpi, 1979a, 1979b; and many others). In geostatistics, the geological phenomenon is considered as a function of space and/or time. Let $Z(x)$ represent any random function of the spatial coordinates z with measured values at N locations in space $z(x_i)$, $i = 1, 2, 3, \ldots, N$, and suppose that the value of the function Z has to be estimated at the point x_o. Each measured value $z_1, z_2, z_3, \ldots, z_N$ contributes in part to the estimation of the unknown value z_o at location x_o. Taking this into account, and assuming a linear relation between z_o and z_i, the estimated value of z_o, can be defined as

$$z^*(x_o) = \lambda_1 z_1 + \lambda_2 z_2 + \lambda_3 z_3 + \lambda_4 z_4 + \cdots \lambda_N z_N$$
$$\text{or } z^*(x_o) = \sum_{i=1}^{N} \lambda_i z(x_i) \tag{5.1}$$

where $z^*(x_o)$ is the estimation of function $Z(x)$ at the point x_o and λ_i are weighting coefficients.

To ensure an unbiased and optimal predictor the following two conditions need to be satisfied:

1. The expectant difference between the estimated value and the true (unknown) value (the expected value of the estimation error) should be zero, i.e.

$$E[z^*(x_o) - z(x_o)] = 0 \tag{5.2}$$

This unbiasness condition is often referred to as the universality condition.

2. The condition of optimality means that the variance of the estimation error should be a minimum, i.e.

$$\sigma_K^2(x_o) = \text{var}[z^*(x_o) - z(x_o)] \text{ is a minimum} \tag{5.3}$$

Substituting equation (5.1) into (5.2) we get

$$\sum_{i=1}^{N} \lambda_i = 1 \tag{5.4}$$

Expanding equation (5.3) then leads to

$$\sigma_K^2(x_o) = \sum_{i=1}^{N} \lambda_i \sum_{j=1}^{N} \lambda_j E[z(x_i)z(x_j)] + E[z^*(x_o)^2] - 2\sum_{i=1}^{N} \lambda_i E[z(x_i)z^*(x_o)]$$

Introducing the formulae of covariance, we get

$$\sigma_K^2(x_o) = \sum_{i=1}^{N} \lambda_i \sum_{j=1}^{N} \lambda_j C(x_i, x_j) + C(0) - 2\sum_{i=1}^{N} \lambda_i C(x_i, x_o) \tag{5.5}$$

where $C(x_i, x_j)$ is the covariance between points x_i and x_j.

The best unbiased linear estimator is the one which minimises $\sigma_K^2(x_o)$ under the constraint of equation (5.4). Thus, introducing Lagrange multipliers and adding the term $2\mu\left(1 - \sum_{i=1}^{N} \lambda_i\right)$, we obtain

$$Q = \sum_{i=1}^{N} \lambda_i \sum_{j=1}^{N} \lambda_j C(x_i, x_j) + C(0) - 2\sum_{i=1}^{N} \lambda_i C(x_i, x_o) + 2\mu - 2\mu\sum_{i=1}^{N} \lambda_i \tag{5.6}$$

Making partial differential equations of Q with respect to λ_i and μ and equating them to zero minimises equation (5.6) leading to the kriging equations:

$$C(x_i, x_o) = -\mu + \sum_{j=1}^{N} \lambda_j C(x_i, x_j), \ i = 1, 2, 3, \ldots, N \tag{5.7}$$

$$\sum_{j=1}^{N} \lambda_j = 1 \tag{5.8}$$

Substituting equation (5.7) into equation (5.5) yields the variance of the estimation error:

$$\sigma_K^2(x_o) = C(0) + \mu - \sum_{i=1}^{N} \lambda_i C(x_i, x_o) \tag{5.9}$$

The square root of this equation gives the standard deviation $\sigma_K(x_o)$, which means that, with 95% confidence, the true value will be within $z^*(x_o) \pm 2\sigma_K(x_o)$.

If the covariance cannot be defined we can apply the kriging equation:

$$\gamma(x_i, x_o) = \mu + \sum_{j=1}^{N} \lambda_j \gamma(x_i, x_j), \ i = 1, 2, 3, \ldots, N \tag{5.10}$$

where $\gamma(x_i, x_j)$ is the variogram between points x_i and x_j (see below). The variance of the estimation error then becomes

$$\sigma_K^2(x_o) = \mu + \sum_{i=1}^{N} \lambda_i \gamma(x_i, x_o) \tag{5.11}$$

Equation (5.10) in conjunction with (5.8) represents a set of $(N + 1)$ linear equations with $(N + 1)$ unknowns, which on solution yield the coefficients λ_i. Once these are known, equations (5.1), (5.9) and/or (5.11) can be evaluated.

Kriging is the best linear unbiased estimator (BLUE) as this estimator is a linear function of the data with weighting coefficients calculated according to the specifications of unbaisedness and minimum variance. The weighting coefficients are determined by solving a system of linear equations with coefficients that depends only on the variogram, which describes the spatial structure (correlation in space) of the function Z. The weights are not selected on the basis of some arbitrary rule, but depend on the statistics concerning how the function varies in space. A major advantage of kriging is that it provides an estimate of the magnitude of the estimation error, which can be used as a rational measure of the reliability of the estimate. The variance of the estimation error depends on the variogram and on the measurement locations. Therefore, before deciding on the selection of a new location for measurement (or deletion of an existing measurement point), the variance of the new estimation error can be calculated. In this way a measurement network can be optimised to produce a minimum variance.

5.2.1 The variogram

It has been observed that all important hydrogeological properties and parameters (piezometric head, transmissivity or hydraulic conductivity, storage coefficient, specific yield, thickness of aquifer, hydrochemical parameters, etc.) are functions of space. These variables (known as regionalised variables) are not purely random. There is generally some kind of correlation in the spatial distribution of their magnitudes (de Marsily, 1986). The spatial correlation of such variables is called the structure, and is normally described by the variogram. The experimental variogram measures the average dissimilarity between data separated by a vector d (Goovaerts, 1997). It is calculated according to

$$\gamma(d) = \frac{1}{2N(d)}\sum_{i=1}^{N(d)}[z(x_i + d) - z(x_i)]^2$$

where d is the separation distance between two points, known as the lag distance.

A generalised formula to calculate the experimental variogram from a set of anisotropic scattered data can be written as follows (Ahmed, 1995):

$$\gamma(\underline{d}, \underline{\theta}) = \frac{1}{2N_d}\sum_{i=1}^{N_d}\left[z(x_i + \hat{d}, \hat{\theta}) - z(x_i, \hat{\theta})\right]^2 \qquad (5.12)$$

where

$$d - \Delta d \leq \hat{d} \leq d + \Delta d, \quad \theta - \Delta\theta \leq \hat{\theta} \leq \theta + \Delta\theta \qquad (5.13)$$

The terms d and θ are the initially chosen lags and directions of the variogram, with Δd and $\Delta\theta$ being their respective tolerances. N_d is the number of pairs for a particular lag and direction.

We now introduce the terms actual lag, $\underline{d}$, and direction, $\underline{\theta}$, for the corresponding calculated variogram. To avoid rounding errors associated with pre-decided lags and directions, we can say

$$\underline{d} = \frac{1}{N_d}\sum_{i=1}^{N_d}\hat{d}_i \quad \text{and} \quad \underline{\theta} = \frac{1}{N_d}\sum_{i=1}^{N_d}\hat{\theta}_i \qquad (5.14)$$

It is important to account for every term carefully while calculating variograms. If the data are collected on a regular grid, Δd and $\Delta\theta$ can be taken as zero and $\underline{d}$ and $\underline{\theta}$ become d and θ, respectively. Often, hydrogeological parameters exhibit anisotropy and hence variograms should be calculated in at least two directions to ensure the existence or absence of anisotropy. However, this frequently leads to an impractical number of samples being required.

The variogram model is often the principal input for both interpolation and simulation schemes. However, modelling the variogram is not a unique process although various studies (e.g. Cushman, 1990; Dagan and Neuman, 1997) have attempted to

mathematically relate commonly used variogram parameters to physical characteristics. Model validation is therefore an important process.

Cross-validation is performed by estimating the random function Z at the points where measurements are available (i.e. at data points) and comparing the estimates with the data (e.g. Ahmed and Gupta, 1989). In this exercise, measured values are removed from the data set one by one when the data point is estimated, and the same is repeated for the entire data set. Thus, for all the measurement points, the true measured value (z), the estimated value (z^*) and the variance of the estimation error (σ^2) become available. Consequently the following criteria must be satisfied:

$$\frac{1}{N}\sum_{i=1}^{N}(z_i - z_i^*) \approx 0.0 \qquad (5.15)$$

$$\frac{1}{N}\sum_{i=1}^{N}(z_i - z_i^*)^2 \approx \min \qquad (5.16)$$

$$\frac{1}{N}\sum_{i=1}^{N}(z_i - z_i^*)^2/\sigma_i^2 = 1 \qquad (5.17)$$

$$\frac{|z_i - z_i^*|}{\sigma_i} \leq 2.0 \quad \forall i \qquad (5.18)$$

Various parameters of the variogram model are gradually modified to obtain satisfactory values from the expressions in equations (5.15) to (5.18). During the cross-validation, many important tasks are accomplished, including:

- testing the validity of the structural model (i.e. the selected variogram);
- deciding the optimum neighbourhood for estimation;
- selecting suitable combinations of additional information, particularly in case of multivariate estimation;
- identifying unreliable data.

5.3 APPLICATION OF GEOSTATISTICS IN GROUNDWATER MODELLING

5.3.1 Theoretical aspects

The formulation of the general groundwater flow equation in porous media is based on mass conservation and Darcy's law, and can be represented in three dimensions as follows (e.g. Bear, 1979; Mercer and Faust, 1981; de Marsily, 1986):

$$\frac{\partial}{\partial x}\left(K_x\frac{\partial h}{\partial x}\right) + \frac{\partial}{\partial y}\left(K_y\frac{\partial h}{\partial y}\right) + \frac{\partial}{\partial z}\left(K_z\frac{\partial h}{\partial z}\right) = \pm R' + S_s\frac{\partial h}{\partial t} \qquad (5.19)$$

where K_x, K_y and K_z [LT^{-1}] are the hydraulic conductivities (or permeability) along the x, y and z coordinates, respectively, assumed to be the principal directions of anisotropy; h [L] is the hydraulic head given by $h = p/\rho g + z$ (p is pressure, ρ is the density of water and g is gravitational acceleration) at an elevation z (i.e. the water level measured in piezometers); R' [T^{-1}] is an external volumetric flux per unit volume entering or exiting the system; and S_s [L^{-1}] is the specific storage coefficient.

For flow in confined aquifers where the vertical hydraulic gradient can be ignored (the Dupuit assumption), equation (5.19) can be integrated to get

$$\frac{\partial}{\partial x}\left(T_x \frac{\partial h}{\partial x}\right) + \frac{\partial}{\partial y}\left(T_y \frac{\partial h}{\partial y}\right) = \pm R + S\frac{\partial h}{\partial t} \qquad (5.20)$$

where T_x and T_y [L^{2}T^{-1}] are the transmissivities (product of the hydraulic conductivity and saturated thickness of the aquifer) in the x and y directions (assumed to be the principal directions of anisotropy in the plane, respectively); R [LT^{-1}] is an external volumetric flux per horizontal unit area entering or exiting the aquifer; and S [–] is the storativity (product of the specific storage coefficient and saturated thickness of the aquifer). In practice, the term R represents recharge, evapotranspiration from a shallow aquifer or a leakage flux between different aquifers.

For unconfined conditions (with the Dupuit assumption) equation (5.20) can be written as

$$\frac{\partial}{\partial x}\left(K_x h \frac{\partial h}{\partial x}\right) + \frac{\partial}{\partial y}\left(K_y h \frac{\partial h}{\partial y}\right) = \pm R + S_y\frac{\partial h}{\partial t} \qquad (5.21)$$

where S_y [–] indicates the specific yield of the water-table aquifer. In this case, the hydraulic head is expressly taken from the bottom of the aquifer (assumed horizontal) as the $z = 0$ reference plane, such that the head h is also the saturated thickness of the aquifer. Providing that the fluctuations in h are small compared to the mean, it is possible to approximate equation (5.21) with equation (5.20) but with $S = S_y$ and transmissivity taken as the product of the hydraulic conductivity and the temporally averaged head.

These equations with the appropriate boundary and initial conditions can be solved by standard numerical methods such as finite differences (see the previous chapter) or finite elements. A large body of literature is available on the analytical or numerical formulation, development and solution of the groundwater flow equations. Some useful references are: Freeze (1971); Remson *et al.* (1971); Prickett (1975); Trescott *et al.* (1976); Pinder and Gray (1977); Cooley (1977, 1979, 1982); Brebbia (1978); Mercer and Faust (1981); Wang and Anderson (1982); Townley and Wilson (1985); de Marsily (1986).

The main steps involved in numerically simulating groundwater flow within an aquifer can be summarised as follows:

1. conceptualisation of the aquifer system and establishment of the various flow components taking place in it;
2. preparing a database of the essential and available parameters with their georeference;

3. analysing the spatial variability of the various parameters in the area;
4. discretisation of the aquifer system into meshes with theoretical and practical constraints;
5. fabrication of the aquifer model by assigning various parameters to each mesh;
6. execution of model under steady and transient conditions and comparing the output with field values;
7. calibration of the aquifer model to eliminate mismatches;
8. performing sensitivity analyses to determine the sensitivity of the outcome to the various parameters;
9. finalisation of the calibrated model simulating the flow in the aquifer system;
10. building futuristic scenarios and predicting the water levels using the calibrated model.

5.3.2 Discretisation

Here the modellers have to descritise the area into a large number of uniform grid blocks or meshes over which it is assumed that there is no variation in aquifer properties. Although with the invention of advanced computers the computation of a large number of meshes is not a difficult task, the preparation of data to be assigned to each grid is time-consuming. The main question arising here is how to decide on the number of grid blocks and their sizes. Here the use of geostatistics can help the modelling task. Initially the area is divided into very fine grids and, with the help of geostatistical estimation procedures, the variance of the estimation error of the measured hydraulic head gradient over each grid block is calculated. As we know that the closer the spatial distance between two points, the closer will be the head gradient values, the final grid size can be decided by merging the grids having the same or close values of the variance of the estimation error of the head gradient. Thus the grid size is decided on the basis of geostatistical estimations of hydraulic head gradient which provide unbiased, minimum variance estimates with a unique value over the entire area of the mesh.

5.3.3 Model fabrication

After discretisation of the area using an optimal grid size, hydrogeological parameters are required for each grid block. It is often impossible to determine these parameters accurately in the field for such a large number of points, either for economic reasons or because of difficulties associated with accessing the area. The data are generally only available at a few places in the area of concern.

Deterministic approaches are often not appropriate for assigning parameter values to grid blocks because of great spatio-temporal variability of hydrogeological parameters. Field heterogeneity of groundwater basins is often inextricable and very difficult to analyse with deterministic methods (e.g. Bakr *et al.*, 1978; Delhomme, 1979; Ganoulis and Morel-Seytoux, 1985). In-situ measurements at the basin scale have shown that physical properties of hydrogeological variables are highly irregular. However, by using geostatistical estimation techniques as shown above, it is not only possible to evaluate the values of the parameters at each and every grid block but also the level of accuracy in terms of the variance of estimation error.

5.3.4 Calibration

Calibration is the process of modifying the input parameters to a groundwater model until the output from the model matches an observed set of data within some acceptable criteria. A confidence interval is given by the standard deviation of the estimation error of each parameter, which can be used as a guide to parameter modification at each mesh and to check that the calculated value of the hydraulic head falls inside the confidence interval of the observed values (also see previous chapter).

In steady-state conditions, an unbiased calibration of the aquifer model is obtained by using the head data kriging estimation variance if the results lie within the 95% of confidence interval. Mathematically this can be stated as

$$\left| \frac{h_i^m - h_i^*}{\sigma_i} \right| < 2 \qquad (5.22)$$

where h_i^m and h_i^* are the modelled and estimated (from the observed) water levels and σ_i is the standard deviation of the water level estimation error.

Under transient conditions, the calibration of the aquifer model in an unbiased way is performed using the same criterion on the head averaged over all measurement time steps. In both cases, the hydraulic conductivity is modified in order to stay within the 95% confidence interval of its estimation variance in each mesh. Mathematically, this can be stated as

$$K^* - 2\sigma_k \leq K^T \leq K^* + 2\sigma_k \qquad (5.23)$$

where K^* is the estimated hydraulic conductivity from kriging for a given mesh, K^T is the calibrated hydraulic conductivity for that mesh and σ_k is the standard deviation of the hydraulic conductivity estimation error.

5.3.5 Prediction

After the model has been successfully calibrated, it is now ready for predictive simulations. Thus the model can be used for predicting future groundwater flow or contaminant transport conditions. Of course, errors and uncertainties in a groundwater flow or solute transport analysis makes any single-model predictions uncertain. Methods for dealing with this are discussed in the previous chapter.

5.4 A CASE STUDY FROM THE MAHESHWARAM WATERSHED, ANDHRA PRADESH, INDIA

The Maheshwaram watershed is an experimental research site of about $53 \, \text{km}^2$ in the Ranga Reddy district of Andhra Pradesh, India (Figure 5.1). This watershed is underlain by granitic rocks and is representative of southern India catchments in terms of over-exploitation of its weathered hard-rock aquifer (more than 700 abstraction wells in use), its cropping pattern, rural socio-economy (based mainly on traditional agriculture), agricultural practices and semi-arid climate.

The granite outcrops in and around Maheshwaram form part of the largest of all granite bodies recorded in peninsular India. Alcaline intrusions, aplite, pegmatite, epidote, quartz veins and dolerite dykes traverse the granite. There are three types of fracture patterns in the area: (i) mineralised fractures, (ii) fractures traversed by dykes, and (iii) late-stage fractures represented by joints. The vertical fracture pattern is partly responsible for the development of the weathered zone and the horizontal fractures are the result of this weathering, as shown by Maréchal *et al.* (2003). Hydrogeologically, the aquifer occurs both in the weathered zone and in the underlying weathered-fractured zone. However, due to deep drilling and heavy groundwater withdrawal, the weathered zone has now become dry.

About 150 dug wells were examined and the nature of the weathering was studied. The weathered-zone profiles range in thickness from 1 to 5 m below ground level (b.g.l.). They are followed by semi-weathered and fractured zones that reach down to 20 m b.g.l. The weathering of the granite has occurred in different phases and the granitic batholith appears to be a composite body that has emerged in different places and not as a single body. One set of pegmatite veins displacing another set of pegmatitic veins has been observed in some well sections. Joints are well developed in the main directions: N0–15E, NE–SW and NW–SE.

The groundwater flow system is local, i.e. with its recharge area at a topographic high and its discharge area at a topographic low adjacent to each other. Intermediate and regional groundwater flow systems also exist owing to significant hydraulic conductivity at depth. Aquifers occur in the permeable saprolite (weathered) layer, as well as in the weathered-fractured zone of the bedrock and the quartz pegmatite intrusive veins where they are jointed and fractured. Thus, only the development of the saprolite zone and the fracturing and interconnectivity between the various fractures allow a potential aquifer to develop, provided that a recharge zone is connected to the groundwater system.

The mean annual precipitation (P) is about 750 mm, of which more than 90% falls during the monsoon season. The mean annual temperature is about 26 °C, although in summer ('Rabi' season from March to May), the maximum temperature can reach 45 °C. The resulting potential evapotranspiration (PET)

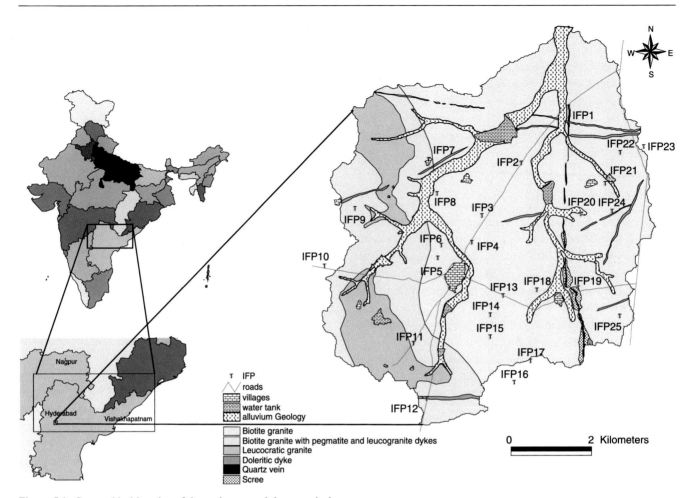

Figure 5.1. Geographical location of the study area and the watershed.

is 1800 mm per year. Therefore, the aridity index (AI = P/PET = 0.42) is within the range 0.2–0.5, typical of semi-arid areas (UNEP, 1992). Although the annual recharge is around 10–15% of rainfall, the water levels have been lowered by about 10 m during the last two decades due to intensive exploitation.

Pumping test analysis from seven wells in the fracture aquifer suggest low storativity ($S \approx 0.006$) and highly variable permeability with transmissivity values ranging from $1.7 \times 10^{-5} \, \mathrm{m^2 \, s^{-1}}$ to $1.7 \times 10^{-3} \, \mathrm{m^2 \, s^{-1}}$ (Maréchal *et al.*, 2004). The groundwater abstraction, which increases year on year, has to be controlled to allow recharge by rainfall to maintain and restore the productive capacity of this depleted aquifer.

The bulk of the permeability is within a fractured-weathered layer located between 20 and 35 m below ground level where fracture zone transmissivities often exceed $5 \times 10^{-6} \, \mathrm{m^2 \, s^{-1}}$ (Maréchal *et al.*, 2004). Below 35 m, the formation is unweathered and much less conductive. To further explore the vertical distribution of hydraulic conductivity, a statistical analysis of airlift flow rates in 288 boreholes, 10 to 90 m deep, was undertaken. Cumulative airlift flow rate were found to range from $1.5 \, \mathrm{m^3 \, h^{-1}}$ to $45 \, \mathrm{m^3 \, h^{-1}}$, with the mean value dramatically increasing in the weathered-fractured layer at depths between 20 and 30 m. Below 30 m, the

flow rate was constant and did not increase with depth. In practice, it is also found that drilling deeper than the bottom of the weathered-fractured layer does not increase the probability of improving the well discharge. The above observations give rise to a conceptual model whereby the bulk of groundwater flow is assumed to occur through a thin weathered-fractured layer. Such a system is common in granitic formations across the world (e.g. Houston and Lewis, 1988; Taylor and Howard, 2000). In the absence of weathering, it is generally observed that permeable features within the granite are limited to poorly connected and sparsely located deep tectonic fractures (e.g. Sweden:Talbot and Sirat, 2001; Finland: Elo, 1992; and the United States: Stuckless and Dudley, 2002). The universal character of granite weathering and its worldwide distribution underline the importance of understanding its impact on the hydro-dynamic properties of the aquifers in these environments.

Two different scales of fracture networks have been identified and characterised by hydraulic tests (Maréchal *et al.*, 2004): a primary fracture network (PFN) which affects the matrix at the decimetre scale, and a secondary fracture network (SFN) affecting the blocks at the borehole scale.

The SFN is composed of a horizontal (HSFN) and sub-vertical fractures (VSFN) set, the latter of which can be clearly observed

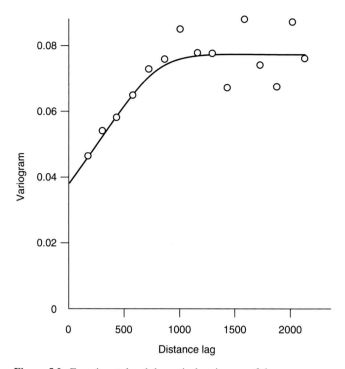

Figure 5.2. Experimental and theoretical variogram of the water level gradient.

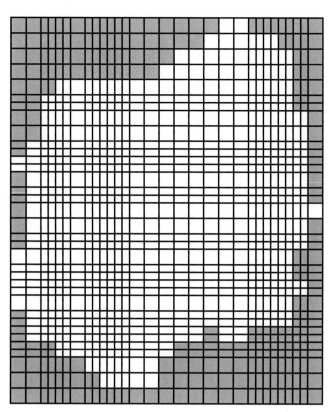

Figure 5.4. Variable size mesh discretisation suitable to MODFLOW.

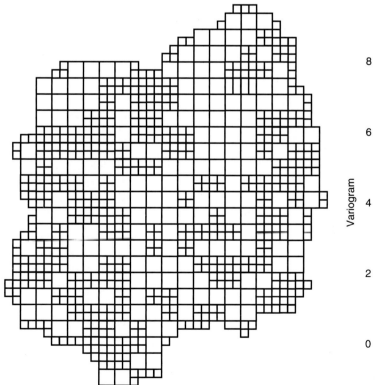

Figure 5.3. The discretisation of the aquifer into nested square meshes.

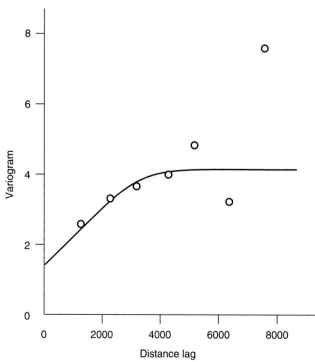

Figure 5.5. Variogram of the percentage of rainfall recharge.

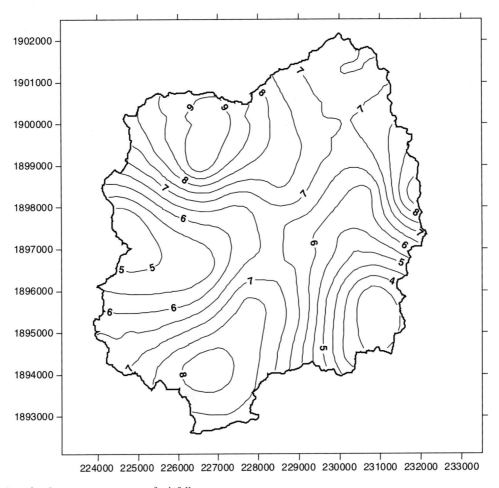

Figure 5.6. Variation of recharge as a percentage of rainfall.

on the outcrops. They are the main contributors to the permeability of the weathered-fractured layer. The average vertical density of the HSFN ranges from $0.15\,\mathrm{m}^{-1}$ to $0.24\,\mathrm{m}^{-1}$, with a fracture length of a few tens of metres (currently available data show a variation between 10 and 30 m). This corresponds to a mean vertical thickness of matrix blocks ranging from around 4 m to 7 m. Strong dependence of the permeability on the density of permeable fractures indicates that individual fractures contribute more or less equally to the bulk horizontal conductivity of the aquifer ($K_r \approx 10^{-5}\mathrm{m\,s}^{-1}$). The VSFN connects the horizontal network, ensuring vertical permeability ($K_z \approx 10^{-6}\mathrm{m\,s}^{-1}$) and good connectivity in the aquifer. Nevertheless, because of the preponderance of horizontal fractures there is a horizontal-to-vertical anisotropy of around 10.

As indicated earlier, the presence of the HSFN is due to weathering processes associated with the inducement of cracks in the rock caused by the expansion of micaceous minerals. These fractures are mostly sub-parallel to the contemporaneous weathering surface, as in the flat Maheshwaram watershed where they are mostly horizontal (Maréchal *et al.*, 2003). This is corroborated by the fact that of the 288 wells drilled in the area,

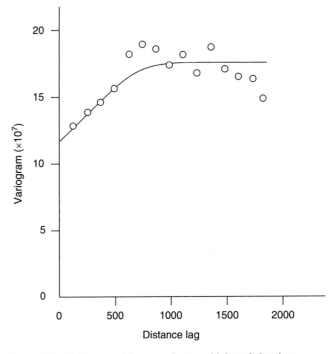

Figure 5.7. Variogram of the groundwater withdrawal showing a high nugget effect.

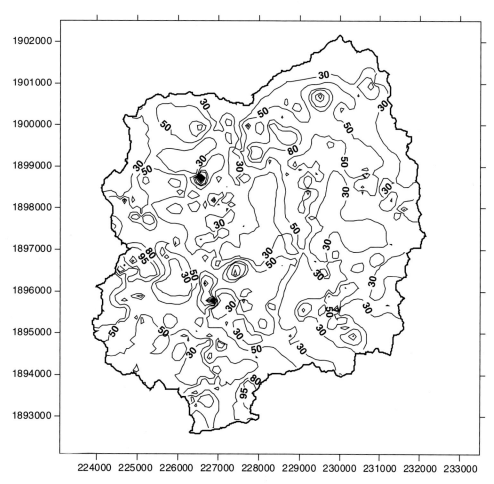

Figure 5.8. Estimated values of the groundwater withdrawals for a typical period.

257 (89%) were drilled deeper than 20 metres and 98% of these are still productive.

The primary fracture network (PFN), operating at the block scale, increases the original matrix permeability (typically $10^{-14}<K_m<10^{-9}$ m s^{-1}) to around $K_b \approx 4 \times 10^{-8}$ m s^{-1}. The PFN also contributes to the matrix block storativity, taking it to around $S_b \approx 5.7 \times 10^{-3}$. This enhanced storage in the blocks represents 91% of the total specific yield ($S_y \approx 6.3 \times 10^{-3}$) of the aquifer; storage in the SFN accounts for the rest. This high storage in the blocks (in both the matrix and the PFN) and the generation of the PFN is believed to result from the first stage of the weathering process. The development of the SFN is associated with a second stage in weathering; hence 'primary' and 'secondary' have been chosen to qualify the different levels of fracture networks. These storage values are consistent with typical unconfined aquifers in low-permeability sedimentary layers.

Several input parameters were analysed for their variability to assess the areas of high, low and medium variability.

Observed water level gradients over the area were kriged to a reasonably fine uniform and isotropic grid of 200 m resolution. Figure 5.2 shows the experimental and theoretical variogram of the water level gradient. Estimation error of the water level gradient in a pre-monsoon season (when pumping is a minimum and water levels are relatively undisturbed by the anthropogenic activity) were taken as a guiding parameter for localised grid resolution reduction by joining the grid blocks with uniform values. This resulted in the 'nested square' grid shown in Figure 5.3.

In this way, discretisation is based on the criteria of parameter variability and distribution of the estimation error. Unfortunately, MODFLOW does not permit a nested square grid. Therefore, the grid needs to be refined again in some areas such that we end up with the system detailed in Figure 5.4. There is software that permits the use of the originally obtained mesh discretisation (e.g. NEWSAM; de Marsily et al., 1978) but generally the USGS MODFLOW package is popular due to the fact that it is free and easily available in simple-to-use windows-based interfaces (e.g. PMWIN; Chiang, 2005). The grid sizes thus used are 200 m by 200 m, 200 m by 400 m, 400 m by 200 m and 400 m by 400 m, with a total of 764 grid blocks in the watershed (Figure 5.4).

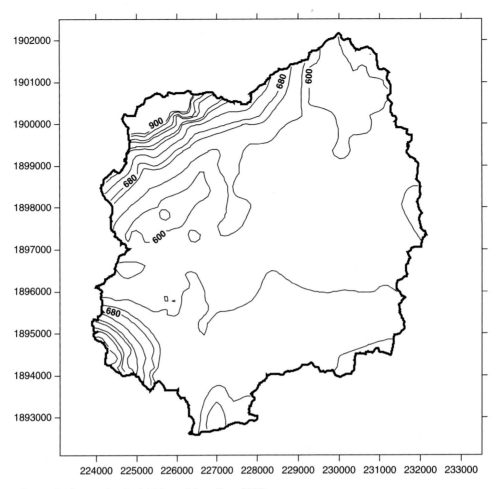

Figure 5.9. Piezometric map in the area for the initial conditions (June 2000).

The various system parameters were then kriged to the 764 grid blocks using variograms constricted for the available data. For hydraulic conductivity this included results from hydraulic testing of 25 boreholes distributed across the area. When kriging hydraulic conductivity it is sensible to log-transform since this parameter is generally log-normally distributed (see the previous chapter). Detailed specific yield data was not available; therefore just two zones were defined. The eastern and western boundaries were taken as no-flow boundaries because the major flow direction is from south to north. The remaining nodes at the boundaries were taken as prescribed head boundaries as the water levels are well known. Other input parameters such as rainfall recharge, groundwater abstraction, groundwater evaporation, irrigation return flow from irrigated fields, etc. were treated as regionalised variables and estimated using kriging. Figures 5.5 and 5.6 show the variogram and kriged maps of recharge percentage respectively. Note that the number of grid blocks and pumping wells are comparable. Therefore, a value of abstraction was also kriged for each grid block. Figures 5.7

and 5.8 show the variogram and estimates of groundwater withdrawals respectively. The model with these parameter values was run under steady and transient conditions. Figure 5.9 shows the estimated values of the initial water level that was used for transient simulation.

The most important aspect in dealing with and removing the bias from the aquifer modelling is during calibration and acceptance of the model. The criterion in equation (5.22) was used to test the acceptability of the calibration output, with changes during calibration based on equation (5.23). The values of equation (5.22) (model calibration error) are plotted in Figure 5.10. Such maps are helpful for identifying areas in the grid that require further calibration.

5.5 CONCLUSIONS

It is often very difficult to analyse field heterogeneities of groundwater basins with deterministic methods. Thus, stochastic

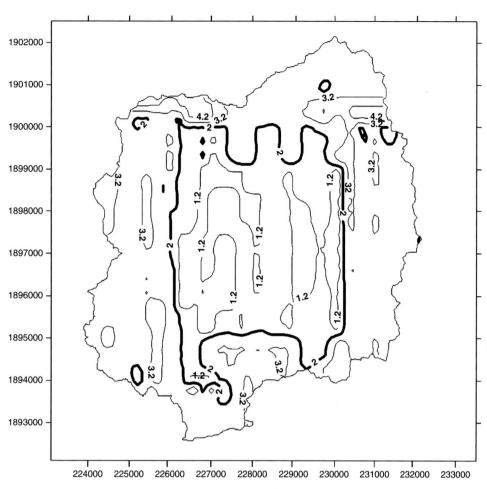

Figure 5.10. Areas that need calibration and areas where the comparison of water levels is satisfactory (inside the thick black curves).

methods have been used in this example especially for the estimation of the hydrogeological parameters. The theory of regionalised variables provides a sound stochastic model for the study of spatial phenomena and its application in the field of hydrogeology has been found to be extremely useful. The advantage of geostatistical methods is that they provide the variance of the estimation error together with the estimated value. Of course, there are tremendous applications of this method particularly in groundwater modelling:

- Better estimated values with lower estimation variance are initially assigned to the nodes of an aquifer model using geostatistical estimation. The closer the values of aquifer parameters to reality, the faster will be the model calibration.

- An assumption is made in aquifer modelling that a single value of the system parameter is assigned to each mesh of the entire grid (although very small). Averaging over a grid block in two or three dimensions can be obtained through block kriging (see e.g. Chiles and Delfiner, 1999).

- An optimal grid size and number of nodes in discretising an aquifer system can be obtained by kriging the head gradients and best locations for the new control points can be defined.

- A confidence interval given by the standard deviation of the estimation error provides a useful guide to parameter modification over each grid block and a check that the calculated heads fall inside the confidence interval of the estimated heads based on the observations.

- A performance analysis of several equivalent calibrated models can be made by comparing the variance of the calibrated head error with the variance of observed head estimation error.

The main limitation in applying geostatistical techniques to hydrogeological problems is due to the scarcity of observed data. Careful estimation of the variogram and cross-validation of the variogram models are therefore essential. However, modifications and improvement to the existing kriging techniques can improve this situation (e.g. the kriging with external drift technique; see Ahmed and de Marsily, 1988, or Chiles and Delfiner, 1999).

REFERENCES

Aboufirassi, M. and Marino, M. A. (1983) Kriging of water levels in Souss aquifer, Morocco. *J. Math. Geol.* **15**(4), 537–551.

Aboufirassi, M. and Marino, M. A. (1984) Cokriging of aquifer transmissivities from field measurements of transmissivity and specific capacity. *J. Math. Geol.* **16**(1), 19–35.

Ahmed, S. (1995) An interactive software for computing and modeling a variogram. In *Proc. of a conference on Water Resources Management (WRM '95), August 28–30*, ed. Mousavi and Karamooz, 797–808. Isfahan University of Technology, Iran.

Ahmed, S. and de Marsily, G. de (1988) Some application of multivariate kriging in groundwater hydrology. *Science de la Terre, Serie Informatique* **28**, 1–25.

Ahmed, S. and Gupta, C. P. (1989) Stochastic spatial prediction of hydrogeologic parameters: role of cross-validation in kriging. In *International Workshop on Appropriate Methodologies for Development and Management of Groundwater Resources in Developing Countries,* Hyderabad, India, February 28–March 4, 1989. Oxford and IBH Publishing Co. Pvt. Ltd, IGW, Vol. III, 77–90.

Armstrong, M. (1998) *Basic Linear Geostatistics*, Springer.

Bakr, A. A., Gelhar, L. W., Gutjahr, A. L. and MacMillan, R. R. (1978) Stochastic analysis of spatial variability in subsurface flows. 1. Comparison of one- and three-dimensional flows. *Water Resour. Res.* **14**(2), 263–271.

Bear, J. (1979) *Hydraulics of Groundwater*. McGraw-Hill.

Brebbia, C. A. (1978) *The Boundary Element Method for Engineers*. Pentech Press.

Chiang, W. H. (2005) *3D-Groundwater Modeling with PMWIN: a simulation system for modeling groundwater flow and transport processes*, 2nd edn. Springer.

Chiles, J. P. and Delfiner, P. (1999) *Geostatistics: modeling spatial uncertainty*. Wiley.

Cooley, R. L. (1977) A method of estimating parameters and assessing reliability for models of steady state groundwater flow. 1. Theory and numerical property. *Water Resour. Res.* **13**(2), 318–324.

Cooley, R. L. (1979) A method of estimating parameters and assessing reliability for models of steady state groundwater flow. 2. Application of statistical analysis. *Water Resour. Res.* **15**(3), 603–617.

Cooley, R. L. (1982) Incorporation of prior information into non-linear regression groundwater flow models. 1. Theory. *Water Resour. Res.* **18**(4), 965–976.

Cressie, N. A. (1991) *Statistics for Spatial Data*. Wiley Series in Probability and Mathematical Statistics. Wiley.

Cushman, J. H. (1990) An Introduction to hierarchial porous media. In *Dynamics of Fluids in Hierarchial porous media*, ed. J. H. Cushman, 1–5. Academic.

Dagan, G. and Neuman, S. P. (1997) *Subsurface flow and transport: a stochastic approach*. Cambridge University Press.

de Marsily, G. (1986) *Quantitative Hydrogeology: Groundwater Hydrology for Engineers*, 203–206. Academic.

de Marsily, G., Ledoux, E., Levassor, A., Poitrinal, D. and Salem, A. (1978) Modelling of large multilayered aquifer systems: theory and applications. *J. of Hydrol.* **36**, 1–33.

de Marsily, G., Lavedan, G., Boucher, M. and Fasanini, G. (1984) Interpretation of interference tests in a well field using geostatistical techniques to fit the permeability distribution in a reservoir model. In *Geostatistics for Natural Resources Characterization,* part 2, ed. G. Verly et al., 831–849. D. Reidel.

Delfiner, P. and Delhomme, J. P. (1983) Optimum interpolation by kriging. In *Display and Analysis of Spatial Data*, ed. J. C. Davis and M. J. McCullough, 96–114. Wiley.

Delhomme, J. P. (1976) *Application de la théorie des variables régionalisées dans les sciences de l'eau*. Doctoral Thesis, Ecole des Mines de Paris Fontainebleau, France.

Delhomme, J. P. (1978) Kriging in the hydrosciences. *Advances in Water Resources* **1**(5), 252–266.

Delhomme, J. P. (1979) Spatial variability and uncertainty in groundwater flow parameters: a geostatistical approach. *Water Resour. Res.* **15**(2), 269–280.

Elo, S. (1992) Geophysical indications of deep fractures in the Narankavaara-Syote and Kandalaksha-Puolanka zones. *Geological Survey of Finland* **13**. 43–50.

Freeze, R. A. (1971) Three-dimensional, transient, saturated-unsaturated flow in a groundwater basin. *Water Resour. Res.* **7**(2), 347–366.

Gambolati, G. and Volpi, G. (1979a) Groundwater contour mapping in Venice by stochastic interpolators. 1. Theory. *Water Resour. Res.* **15**(2), 281–290.

Gambolati, G. and Volpi, G. (1979b) A conceptual deterministic analysis of the kriging technique in hydrology. *Water Resour. Res.* **15**(3), 625–629.

Ganoulis, J. and Morel-Seytoux, H. (1985) *Application of stochastic methods to the study of aquifer systems*. Project IHP-II A.1.9.2. UNESCO.

Goovaerts, P. (1997) *Geostatistics for Natural Resources Evaluation*. Oxford University Press.

Houston, J. F. T. and Lewis, R. T. (1988) The Victoria Province drought relief project. II. Borehole yield relationships. *Groundwater* **26**(4), 418–426.

Maréchal, J. C., Dewandel, B. and Subrahmanyam, K. (2004) Use of hydraulic tests at different scales to characterise fracture network properties in the weathered-fractured layer of a hard rock aquifer. *Water Resour. Res.* **40**, W11508, doi:10.1029/2004WR003137

Maréchal, J. C., Wyns, R., Lachassagne, P., Subrahmanyam, K. and Touchard, F. (2003) Anisotropie verticale de la perméabilité de l'horizon fissuré des aquifères de socle: concordance avec la structure géologique des profils d'altération, C.R. *Geoscience* **335**. 451–460.

Matheron, G. (1965) *Les variables régionalisées et leur estimation*. Masson.

Mercer, J. W. and Faust, Ch. R. (1981) *Groundwater Modelling*. National Water Well Assoc.

Pinder, G. F. and Gray, W. G. (1977) *Finite Element Simulation in Surface and Subsurface Hydrology*. Academic.

Prickett, T. A. (1975) Modeling techniques for groundwater evaluation. *Adv. Hydrosci.* **10**, 1–143.

Remson, I., Hornberger, G. M. and Molz, F. J. (1971) *Numerical Methods in Subsurface Hydrology with an Introduction to the Finite Element Method*. Wiley (Intersciences).

Stuckless, J. S. and Dudley, W. W. (2002) The geohydrologic setting of Yucca Mountain, Nevada. *Applied Geochemistry* **17**(6), 659–682.

Talbot, C. J. and Sirat, M. (2001) Stress control of hydraulic conductivity in fracture-saturated Swedish bedrock. *Engineering Geology* **61**(2–3), 145–153.

Taylor, R. and Howard, K. (2000) A tectono-geomorphic model of the hydrogeology of deeply weathered crystalline rock: evidence from Uganda. *Hydrogeology J.* **8**(3), 279–294.

Townley, L. R. and Wilson, J. L. (1985) Computationally efficient algorithms for parameter estimation and uncertainty propagation in numerical models of groundwater flow. *Water Resour. Res.* **21**(12), 1851–1860.

Trescott, P. C., Pinder, G. F. and Carson, S. P. (1976) Finite difference model for aquifer simulation in two dimensions with results of numerical experiments. In *Technique of Water Resources Investigations of the USGS*.

United Nations Environmental Programme (1992) *World Atlas of Desertification*. Edward Arnold.

Wang, H. F. and Anderson, M. P. (1982) *Introduction to Groundwater Modeling. Finite Difference and Finite Element Methods*. Freeman.

6 Groundwater vulnerability and protection

A. P. Butler

6.1 INTRODUCTION

Groundwater is a particularly vital resource in arid and semi-arid areas owing to the scarcity of suitable surface-water resources and the high evaporation from surface-water storage. However, if it is also to be a reliable resource, able to meet current and future demand, it has to be managed effectively, not only in terms of quantity (i.e. abstraction) but also quality. Recent decades have seen substantial progress in the development of methods for remediating contaminated groundwater (Reddy, 2008). However, because of the complexities associated with the removal and/or destruction of pollutants in the subsurface this is often both costly and time consuming. The maxim 'prevention is better than cure' is therefore an important one in groundwater resource management.

One of the key instruments for seeking to maintain good groundwater quality is groundwater protection policy. In most developed countries groundwater protection has been formally incorporated into legislation as a means to ensure good groundwater quality on a sustainable basis and as part of an integrated approach to environmental protection (e.g. EU, 2009). In practice, this is achieved through aquifer vulnerability mapping at the regional scale and source zone protection at the scale of local abstractions. This chapter summarises these two complementary approaches to groundwater protection and considers how they can be implemented in arid and semi-arid regions. This focus on climate is important, as the geological and hydrological conditions are generally very different from those in more temperate regions (Robins et al., 2007). Furthermore, the available data required to implement these measures are generally limited and of varying quality, and therefore the methods are often subject to large uncertainty.

6.2 AQUIFER VULNERABILITY

6.2.1 Introduction

All groundwater is vulnerable to pollution from surface activities (Figure 6.1). However, the degree of vulnerability is dependent on a number of factors. Of prime importance is the nature of the groundwater body, i.e. its ability to act as a potential or actual water resource. Rocks suitable for water supply are usually classified as either major or minor aquifers. Major aquifers are extensive, high-yielding water bodies that can be used for regional water supplies. One such example is the Nubian Sandstone (Dawoud, 2004). Covering an area of around 2 million square kilometres, this aquifer is a vital water resource for Chad, Libya, Sudan and Egypt. Minor aquifers are permeable formations whose storage and yield are sufficient for more local requirements, e.g. small towns, villages or individual domestic dwellings (Chilton and Foster, 1995). Formations not capable of yielding sufficient amounts of water for supply are termed non-aquifers (or aquicludes) and generally given a lower priority in terms of protection.

The next key factor is whether there exists a potential flow pathway between the surface and the underlying groundwater body. This depends on the hydrogeological properties of any strata overlying the aquifer and their control on groundwater recharge. For example, permeable rock formations that have one or more overlying impermeable layers are, effectively, isolated from the surface and therefore have a low vulnerability. Conversely, aquifers that outcrop at the surface, or are overlain by permeable or semi-permeable formations that allow infiltrating surface water to percolate down to a water table, are liable to become polluted from any contaminants present in infiltrating water.

The presence of a permeable pathway from the surface to an underlying aquifer does not automatically mean that it is highly vulnerable to pollution. A further factor is the nature of the hydrological processes that give rise to recharge. Robins (1998) has stated that recharge is the key to both groundwater pollution and aquifer vulnerability. This is particularly important in the context of arid and semi-arid environments where rainfall is a highly episodic process, with marked spatial variability of hydrological response (see Chapter 2). It may not give rise to any substantial recharge in some areas (where rainfall that falls is subsequently returned to the atmosphere through

Groundwater Modelling in Arid and Semi-Arid Areas, ed. Howard S. Wheater, Simon A. Mathias and Xin Li. Published by Cambridge University Press.
© Cambridge University Press 2010.

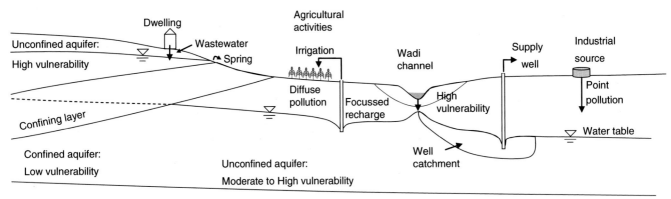

Figure 6.1. Various pollution sources and types of aquifer vulnerability.

evaporation), but it can result in very high levels of recharge, due to flow focusing (such as occurs along wadi channels), at other locations. However, it is important to recognise that, in addition to natural recharge, it is also possible that groundwater contamination can occur from localised releases of fluid in relatively close proximity to the water table (e.g. releases from septic tanks or wastewater collection systems; Simmers, 1998). In this case the driving head of the effluent could be sufficient to pollute an underlying aquifer without any assistance from natural recharge processes. A further mechanism that can result in a near-surface aquifer becoming polluted is when contaminants are artificially mobilised by irrigation.

The final factor affecting aquifer vulnerability is the effect of the overlying material on contaminant transport, as this can result in the attenuation of contaminant concentrations, thereby reducing their long-term impact on groundwater quality. This includes material properties, such as cation exchange capacity, and the presence of micro-organisms, which can degrade organic pollutants and remove nitrates under appropriate conditions, and their vertical extent.

The assessment of an aquifer's vulnerability provides a comparative indicator for determining areas where groundwater is more susceptible to becoming polluted. Methods used to determine this indicator seek to integrate the various factors outlined above into either a single number, or index, or as one of a limited number of categories. Such assessments provide planners and environmental managers with a means of prioritising activities in order to ensure the long-term sustainability of groundwater quality. However, with all aquifer vulnerability assessment methods the following assertions should also be borne in mind (Chilton, 2006; after Foster and Hirata, 1998, and NRC, 1993):

1. All aquifers are vulnerable to persistent, mobile pollutants in the long term.
2. Uncertainty is inherent in all pollution vulnerability assessments.

3. If complex assessment systems are developed, obvious factors may be obscured, and subtle differences may become indistinguishable.

The areal extent of aquifers means that aquifer vulnerability is generally presented in mapped form using geographical information system (GIS) techniques (Hiscock et al., 1995). There are two distinct approaches (Vrba and Zaporozec, 1994): a hierarchical system, based on a limited number of classes (such as mapped areas on solid geology and surficial soils), and parametric methods, which seek to integrate a number of characteristics in a quantitative or semi-quantitative manner. Of the latter, two distinct approaches are generally used: matrix systems (or zoning) and indexing (or rating/point counting; Chilton, 2006). In the case of matrix systems, different aquifer characteristics, combined with the properties of the overlying soil and rock, are used to denote regions of high, low and various intervening levels of vulnerability. By contrast, the indexing method seeks to produce a quantitative measure of vulnerability through a numerical index (see Table 6.1 and also Gogu and Dassargues, 2000, for a description of those in common usage). However, the degree of vulnerability may be subsequently categorised through the use of specified ranges in order to simplify the final representation (Doerfliger et al., 1999; Foster, 1987). Intrinsic vulnerability refers to a generic study (i.e. without reference to a particular contaminant) whereas specific vulnerability refers to a particular contaminant, e.g. nitrate. In the case of the latter, comparisons of specific vulnerabilities with observed groundwater contaminant concentrations have provided a means for assessing the performance of the methodology (Debernardi et al., 2008; Panagopoulos et al., 2006). As well as chemical data, in recent years land use data derived from aerial (Al-Adamat et al., 2003; Denny et al., 2007; Stigter, 2006) and satellite imaging (Dimitriou and Zacharias, 2006) have also been incorporated into groundwater vulnerability studies. This can be particularly useful in developing countries where detailed and extensive maps of land use are not readily available (Werz and Hötzl, 2007).

Table 6.1. *Main indexing methods used for aquifer vulnerability mapping.*

Acronym	Description	Reference
AVI	The aquifer vulnerability index (AVI) is the sum of the hydraulic resistance (thickness/hydraulic conductivity) for each sedimentary layer above the uppermost saturated aquifer. It does not account for other factors such as recharge or topography.	(van Stemport *et al.*, 1993)
DRASTIC	A weighted index method developed by the US EPA. The method utilises seven parameters: depth to groundwater table (D), net recharge (R), aquifer media (A), soil media (S), topography (T), impact of the vadose zone media (I), hydraulic conductivity of the aquifer (C). It has been widely applied in many countries and regions outside the USA.	(Aller *et al.*, 1987)
EPIK	Developed at the University of Neuchatel, Switzerland, for risks posed to groundwater quality in karst systems. Uses four parameters: epikarst (E), protective cover (P), infiltration conditions (I) and karst development (K). These are combined using a weighting system to produce an intrinsic vulnerability map divided into four vulnerability classes.	(Doerfliger *et al.*, 1999)
GLA	Developed by the German State Geological Surveys (Geologische Landesämter; GLA) and the Federal Institute of Geosciences and Natural Resources (Bundesanstalt fuer Geowissenschaften und Rohstoffe; BGR). Evaluates vulnerability by estimation of the transit time of the percolation water using three factors: the thickness of each layer in the unsaturated zone, the permeability of each stratum of the unsaturated zone, and the amount of percolation water. The evaluation process is for the soil zone and the remaining part of the unsaturated zone. The index ranges from 0 to >4000 and is theoretically related to the percolation time of water through the unsaturated zone.	(Hölting *et al.*, 1995)
GOD	Employs three parameters: nature of groundwater occurrence (G), overall aquifer class (O), which considers the degree of consolidation and lithological character, and depth to groundwater (D), which takes into consideration the confined or unconfined nature of the aquifer. A numerical value is attached to each parameter division and the three values are then multiplied together to form an aquifer vulnerability index.	(Foster, 1987)
PI	Protective cover (P) and infiltration (I) condition. Based on the GLA method but with special consideration of karst-specific behaviour in terms of infiltration conditions. Evaluates protective cover by considering all layers between the ground surface and the water table using the thickness and hydraulic conductivity of each and assessing of the degree of bypassing of the protective cover by surface and near-surface flow (I) which particularly occurs within the catchment of sinking streams. The resulting vulnerability map is calculated by multiplying the P and I factors and shows the spatial distribution of the protective function. The higher the PI factor, the higher the protective function and hence the lower the vulnerability.	(Goldscheider *et al.*, 2000)
SINTACS	This method utilises the same parameters as DRASTIC, but has five different weighting systems depending on the hydrogeological setting. The weighting systems reflect the relative importance of the parameters within different settings, which are known as Normal, Severe, Seepage, Karst and Fissured.	(Civita and de Maio, 2000)

6.2.2 Indexing methods

INTRINSIC VULNERABILITY

One of the earliest indexing methods was the GOD scheme proposed by Foster (1987). This incorporated three factors, summarised as *Groundwater occurrence including recharge, Overall lithology* and *Depth to groundwater*. Each factor – namely the aquifer type (e.g. confined, semi-confined, unconfined), the lithology of the overlying aquitard or aquiclude (in the case of a confined or semi-confined aquifer) or the aquifer unsaturated zone (in the case of an unconfined aquifer) and the depth to water – was rated as a value between 0 (not vulnerable) to 1 (highly vulnerable). A schematic of the method is shown in

Figure 6.2. The product of these three values, also a value between 0 and 1, gives the overall aquifer pollution vulnerability. A result of less than 0.3 is considered to represent low vulnerability, between 0.3 and 0.5 moderate and a value above 0.5 is considered to be high or, above 0.7, extreme. The method has not been widely used, although its performance in relation to observed nitrate concentrations was recently assessed by Debernardi *et al.* (2008).

At roughly the same time Foster was developing his method in the UK, an alternative and more complex method was being developed by the US EPA (Aller *et al.*, 1987). Known by the acronym DRASTIC, this incorporates seven characteristics of the surface and subsurface system in order to produce a single

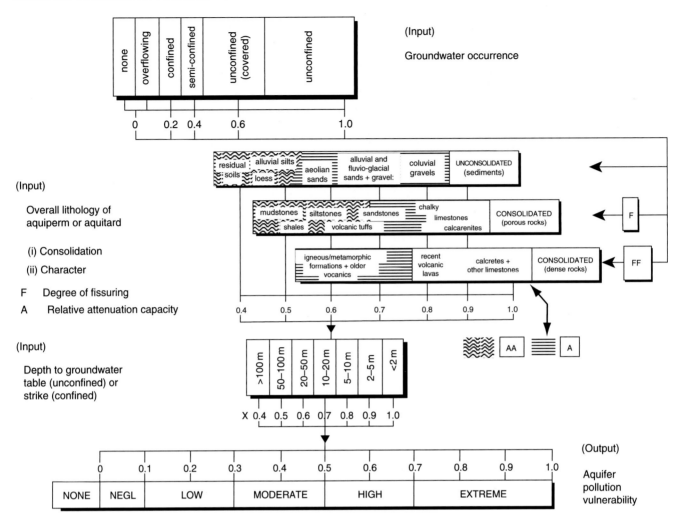

Figure 6.2. GOD system for evaluation of aquifer pollution vulnerability index (after Foster, 1987).

vulnerability index. Each characteristic, or descriptor, is given a value between 1 and 10 and is assigned an associated weight (see Table 6.2 for details). Summing the weighted values allows an overall index value characterising the pollution potential at a given location to be calculated. The vulnerability index has a minimum and maximum value of 23 and 230, respectively. However, this value should be treated in relative rather than absolute terms. As the methodology was originally developed in the USA, the weightings reflect the conditions of that country. However, the diversity of geology, soils and climate across the continent (which also includes the state of Hawaii) has helped to give the system a wider basis. One of the attractions of the method is that it is well suited for use within a geographical information system (GIS) in order to utilise spatial input data and provide a map of the resulting index values (e.g. Chitsazan and Akhtari, 2009; Evans and Myers, 1990; Hamza et al., 2007). However, criticisms have been raised that the method is over-parameterised (Barber et al., 1993; McLay et al., 2001; Merchant, 1994).

Table 6.2. *Descriptors and associated weightings used in DRASTIC.*

Indicator	Description	Weight
D	Depth to water table	5
R	Net Recharge	4
A	Aquifer media	3
S	Soil media	2
T	Topography (slope)	1
I	Impact of the vadose zone	5
C	(Hydraulic) Conductivity of the aquifer	3
	Total	23

In succession to DRASTIC, a number of other indexing methods have also been developed (see Table 6.1). For example, the SINTACS methodology (Civita and de Maio, 2000) is similar to that used in DRASTIC but includes five different weightings dependent on geological conditions.

An alternative, and arguably more physically based, approach to quantifying aquifer vulnerability is based on estimates of the travel time for infiltrating water to reach the water table, the inverse of which is used to denote aquifer vulnerability (i.e. the shorter the travel time, the more easily groundwater can become polluted). In the aquifer vulnerability method (AVI) this is derived solely from the total hydraulic resistance of the various lithological layers between the surface and the saturated zone (van Stemport et al., 1993).

In contrast, the GLA method, developed in Germany (Hölting et al., 1995), includes the amount of infiltrating water. The PI (protective cover/infiltration) method (Goldscheider et al., 2000) is based on the GLA approach but includes a special consideration of karst behaviour that allows for bypassing of the protective cover by surface and near-surface flow, which occurs, for example, with sinking streams. Groundwater in karstic environments is particularly vulnerable and therefore the EPIK method (Doerfliger et al., 1999) was specifically developed to provide vulnerability indices for landscapes which are dominated by carbonate rocks.

Only a few comparative studies using a number of the above approaches have been undertaken and with somewhat mixed results. In a study undertaken for a karst area in southwest Germany (Neukum et al., 2008), vulnerability estimates using four of the above methods were compared with mean unsaturated zone transit times obtained from numerical model simulations using measured hydraulic and transport parameters of the geological sequence. The results showed that the highest level of agreement with the model simulations was achieved with the GLA and PI methods. The reason for the poorer performance of the DRASTIC and EPIK methods was that they are not able to incorporate highly variable distributions and thickness of cover sediments and their protective properties in their respective mapping procedures. A similar type of study was undertaken by Vías et al. (2005) comparing the AVI, GOD, DRASTIC and EPIK methods for carbonate aquifers in southern Spain. They conclude that the GOD method was the most appropriate for vulnerability mapping diffuse flow in such aquifers.

Although the percolation time approach appears to make the vulnerability assessment more objective in comparison to the parameter and weight approaches, which rely on subjective estimates for these values, it has been criticised on the basis that the calculated time-scales are inconsistent with environmental considerations (Andersen and Gosk, 1989). Furthermore, given the difficulties associated with quantifying recharge in arid and semi-arid environments, it can be argued that such an approach might be highly problematic when applied to such regions. This appears to be supported by the fact that very few vulnerability studies in such environments have used this approach (see Section 6.2.4 below).

SPECIFIC VULNERABILITY

Specific vulnerability considers the ability for a particular compound released at or near the surface to pollute an underlying aquifer. It is clearly related to the intrinsic vulnerability. However, whereas the latter takes a conservative view that does not consider the physical and chemical properties of a particular chemical (or group of chemicals), specific vulnerability seeks to account for these. Although, in principle, the specific vulnerability of any chemical can be assessed, in practice it is normally only evaluated for particular (and generally common) pollutants; examples include nitrate (Al-Adamat et al., 2003; Al-Hanbali and Kondoh, 2008; Stigter, 2006), pesticides (Al-Zabet, 2002; Dixon, 2005; Murray and McCray, 2005) and BTEX (Wang et al., 2007).

One advantage of focusing on relatively common pollutants is that comparisons of specific vulnerability for an aquifer pollutant with observed concentrations in groundwater can be used as a means for assessing the performance of the method (Rupert, 2001). However, this is not always a straightforward procedure and the constraint of applying a generic method to a specific contaminant for a particular locality or region often reveals difficulties with the selected method. For example, when investigating and comparing the impact of organic contaminants on groundwater in the city of Wuhan, China, Wang et al. (2007) had to modify the DRASTIC method by ignoring the effects of topography and replaced the soil media and aquifer hydraulic conductivity components with aquifer thickness and a new factor which they termed contaminant impact. Modified versions of the method have also been developed which include for land use (Secunda et al., 1998) and replacing net recharge with recharge potential (Piscopo, 2001).

6.2.3 Matrix methods

Although Foster was one of the originators of the indexing method for assessing groundwater vulnerability (Foster, 1987), he subsequently moved away from this approach and instead advocated a simpler matrix-based method, which relied on the lithology and thickness of strata above a saturated aquifer as the prime indicators of pollution vulnerability (Adams and Foster, 1992). This approach was subsequently adopted by the UK as the national approach for groundwater vulnerability (NRA, 1992) and has also been applied in Botswana (Alemaw et al., 2004) as part of a wellfield protection strategy.

The method has two components: the geological formation being assessed, and the overlying soil and strata. Geological conditions were divided into four classes depending on permeability and the degree of fracturing (reflecting the importance of fractured rocks for the UK's groundwater resources). The effect of the surficial soils and strata on vulnerability was categorised in three classes according to their leaching capability. A schematic of the method is shown in Figure 6.3.

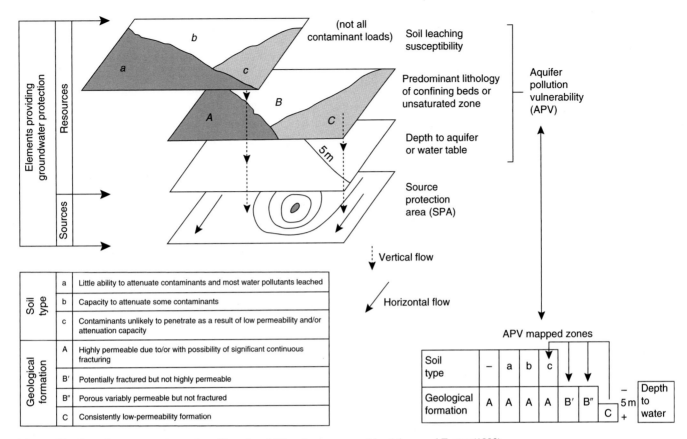

Figure 6.3. Groundwater protection and aquifer vulnerability scheme proposed by Adams and Foster (1992).

Soil type	a	Little ability to attenuate contaminants and most water pollutants leached
	b	Capacity to attenuate some contaminants
	c	Contaminants unlikely to penetrate as a result of low permeability and/or attenuation capacity
Geological formation	A	Highly permeable due to/or with possibility of significant continuous fracturing
	B'	Potentially fractured but not highly permeable
	B"	Porous variably permeable but not fractured
	C	Consistently low-permeability formation

Foster's reason for moving to a matrix (or zoning) method was that 'on balance it was considered preferable not to use indexation, because of the danger of obscuring the basic hydrogeological data on which estimates are based' (Foster and Skinner, 1995). Foster and Skinner (1995) were particularly critical of the DRASTIC indexing method as they considered that it produces a vulnerability index whose meaning is obscure and whose significance is unclear. However, such a view has not been widely shared by the hydrogeological community and the majority of aquifer vulnerability studies are undertaken using indexing methods, the majority of which are based on the original or a modified form of DRASTIC.

6.2.4   Applications of vulnerability methods in arid and semi-arid regions

Attempts to apply vulnerability methods to arid and semi-arid regions have proved challenging due to difficulties in calculating values for some of the descriptors, in particular recharge (Hamza *et al.*, 2007) and the role of the unsaturated zone. In an application of the DRASTIC method to the Paluxy aquifer in north central Texas, Fritch *et al.* (2000) introduced modifications to both of these characteristics. In the case of net recharge, as its calculation at a specific location is problematic given the nature of rainfall

and runoff in semi-arid environments, a value representative of the entire aquifer outcrop was obtained using the recession curve displacement method of Arnold *et al.* (1995). In some of the outcrop areas of the Paluxy aquifer, the unconfined sandstone is overlain by an inter-bedded limestone and shale formation. Weathering of this formation in the near surface has resulted in vertical fracturing, which can enhance water (and hence contaminant) movement in the vadose zone. Therefore, because of its effect on contaminant migration, a methodology to account for the magnitude of this effect in relation to the degree and depth of weathering was incorporated in the DRASTIC index calculation.

As vadose zone impact and depth to the water table have the highest weighting factors, changes in the numerical value of these coefficients can have a marked impact on the overall index value. Furthermore, both limited and localised data (e.g. borehole logs) can present difficulties when interpolating values derived from these sites over large areas. However, improvements in the estimated spatial distribution of water table depth can be achieved through the use of groundwater modelling. As an example, Fredrick *et al.* (2004) used results from an analytical element model. An important aspect of this approach is that it allows the integration of various parameters used to evaluate vulnerability (e.g. aquifer hydraulic conductivity and recharge) to be made in a consistent manner.

Aquifer vulnerability assessments have been conducted in a variety of arid and semi-arid regions including Iran (Chitsazan and Akhtari, 2009), Oman (Jamrah *et al.*, 2008), Tunisia (Hamza *et al.*, 2007), the United Arab Emirates (Al-Zabet, 2002) and the Gaza Strip (Almasri, 2008).

In order to highlight the challenges in undertaking such studies the following discussion focuses on Jordan, where water consumption already exceeds renewable freshwater resources by more than 20% and where freshwater resources are effectively fully utilised, with over 50% of supply derived from groundwater (Al-Adamat *et al.*, 2003). A particular focus has been on nitrate pollution due to intensive agriculture. Al-Adamat *et al.* (2003) used a modified form of the DRASTIC method where the recharge parameter is replaced by the potential of an area to have recharge based on the rainfall amount, slope and soil permeability, as suggested by Piscopo (2001). Model performance was assessed by comparing groundwater nitrate concentrations with the associated vulnerability class at the well location. Results showed a reasonable correlation between the two, albeit with some anomalous results. A similar study was also undertaken by Al Kuisi *et al.* (2006) using the SINTACS model (Civita and de Maio, 2000) to determine areas where groundwater is most vulnerable to contamination.

Aquifer vulnerability assessments in arid and semi-arid regions are particularly important in populated areas where groundwater is a vital, and frequently sole, water resource and where human activity levels mean that there is a high risk of pollution occurring. Al-Hanbali and Kondoh (2008) therefore undertook a groundwater vulnerability assessment within the Dead Sea groundwater basin which coupled DRASTIC with an evaluation of human activity impact (HAI). Due to the geology of the area the evaluation of the net recharge parameter included the effect of intersection between faults and the drainage system. To evaluate human activity, Advanced Spaceborne Thermal Emission and Reflection Radiometer (ASTER) and visible and near infrared (VNIR) data were employed to map urban and agricultural areas within the study area. A comparison of the DRASTIC and HAI indices highlighted areas where human activity was likely to be affecting groundwater quality. Once again the performance of the method was assessed using measured nitrate concentrations in groundwater. A sensitivity study showed that the vadose zone, aquifer media, and recharge parameters had a significant impact on the DRASTIC model, whereas the depth to water table and hydraulic conductivity parameters had no significant effect.

Due to outcrops of karstic limestone in northern Jordan, groundwater in these aquifers is particularly vulnerable to pollution, and studies in these areas have shown groundwater to be microbiologically contaminated as a result of agricultural activities and the disposal of untreated wastewater (Werz and Hötzl, 2007). Assessments of aquifer vulnerability in these areas

have focused on methods incorporating travel time estimates. For example, Chilton (2006) presents a study of the Irbid area, which builds upon a previous study by Margane *et al.* (1999), and uses the GLA method proposed by Hölting *et al.* (1995). The results show that in the main wadis and in the areas where groundwater is close to the ground surface, the vulnerability of the groundwater is extremely high, whereas vulnerability is classified as only moderate on the high plateaux between the deeply incised wadis. Vulnerability is especially high in areas where an effective soil cover is missing, the water table is comparatively shallow and the aquifer is unconfined. Other studies have shown that the zones of highest vulnerability lie within valleys and nearby main fault zones, and that these regions, categorised as protected areas by other methods due to thick unsaturated zones, contribute to a major degree to the total risk (Brosig *et al.*, 2008).

While there has been a growing number of (increasingly sophisticated) vulnerability assessments in semi-arid and arid zone countries, the experience of those undertaking such studies in Africa is pertinent for situations where data and resources are scarce. Robins *et al.* (2007) drew comparisons with the application of vulnerability methods used in developed countries (e.g. the UK) to those in developing countries (e.g. parts of Africa). They conclude that efforts to transfer such methods to Africa have so far met with little success. One reason is that the widespread presence of weathered basement aquifers and their high degree of scale variation means that groundwater vulnerability needs to be assessed at the field scale rather than the regional scale. They also conclude that any vulnerability assessment should exclude aquifer recharge potential (or productivity) to ensure that poorly productive, but socially important, aquifers can be assessed without being overly influenced by reliance on a long-term effective (and often highly uncertain) rainfall value. They suggest that if rating methods such as DRASTIC are used then the weighting of recharge should be reduced. They highlight the importance of protecting individual borehole supplies particularly with respect to the siting of pit latrines and waste disposal sites, where minimum distances between boreholes and point pollution sources of between 15 and 50 m have been recommended (Xu and Braune, 1995). Such protection is a component part of groundwater vulnerability and is generally achieved through the designation of source protection zones.

6.3 SOURCE PROTECTION

6.3.1 Definition

Source protection aims to delineate zones on a map that represent the areas over which a groundwater resource (i.e. well, borehole or spring) receives recharging water from the surface, either from direct rainfall or from focused recharge due to surface

Table 6.3. *Terms and definitions commonly used in source protection zones.*

Term	Definition
Capture zone	The area around the well from which water is captured within a certain time t. Sometimes the time-dependent aspect is emphasised by using the term **time-related capture zone**, but in this chapter the descriptor 'capture zone' will automatically imply time dependence.
Well catchment	The capture zone for infinite time. It encompasses the entire area over which the well draws the water pumped from it. Another term that can refer to both a capture zone and a well catchment is the zone of contribution (ZOC).
Isochrone	The boundary of a capture zone. It is a contour line of equal travel time to the well.
Well head protection area (WHPA)	The area around a pumping well which is associated with a specified level of protection. Its delineation is a political consideration based on a risk analysis, which may or may not involve an actual capture zone delineation.

runoff (Chave *et al.*, 2006). Two types of zones are generally identified: total catchment area of the well, and time-related capture zones. Table 6.3 provides the definitions of various commonly employed terms.

6.3.2 Evaluation

Well capture zones for a specified groundwater residence time are determined by evaluating the corresponding isochrones, which are contour lines of equal travel time from the location within the aquifer to the well. These isochrones can be calculated analytically when pumping at a constant rate within a uniform regional flow field (Bear and Jacobs, 1965). They can also be computed numerically for arbitrarily shaped flow fields (Kinzelbach *et al.*, 1992), using particle tracking techniques. This requires the evaluation of the detailed velocity field and it assumes that advection is the dominant transport process. However, it is also important to recognise that such an evaluation will be affected by uncertainty (Evers and Lerner, 1998; van Leeuwen *et al.*, 1998; Wheater *et al.*, 2000).

For the evaluation of the velocity field, the following parameters and conditions are generally required (Stauffer *et al.*, 2005):

1. Flow geometry: Obtained from hydrogeological investigations, the prevailing flow field can often be approximated by a horizontal two-dimensional (2D) flow and transport model. However, it is important recognise that the extent of the flow domain is subject to uncertainty, mainly due to the extrapolation and interpolation of data.

2. Well pumping rate: This should include the given or planned schedule of the pump. This is frequently the less uncertain of all information. Often, long-term averaged pumping rates can be used. However, the pumping schedule can affect the time-dependent velocity field and hence the capture mechanism of the well.

3. Groundwater recharge rate: This is estimated on the basis of hydrological considerations. Recharge rate can, in general, only be indirectly determined. It is a time- and space-dependent variable. Often these effects cannot be assessed precisely. As noted earlier, even the temporally and spatially averaged recharge rate may show considerable uncertainty, particularly in arid and semi-arid environments (see Chapter 2).

4. Infiltration rate from rivers, creeks and wadis: These can be estimated on the basis of hydrological considerations, or by calibration of a flow model using nearby head and/or concentration data. Infiltration rate is particularly difficult to assess as, in general, it cannot be measured directly and is strongly dependent on local infiltration conditions, which, in turn, can be affected by clogging. Again, rates are, in general, time-dependent and spatially variable.

5. Base and top elevations of the aquifer formation: This information is generally obtained from borehole and/or geophysical investigations. Although generally accurately known at the point of measurement, these are subject to uncertainly when interpolation is involved. In some instances the location of the aquifer does not coincide with the base of the aquifer (e.g. in formations with marked trends in vertical heterogeneity that restrict flows at depth; Williams *et al.*, 2006).

6. Piezometric head of the aquifer: This information is generally obtained from boreholes and/or geophysical investigations. The piezometric head of the aquifer is based on local borehole measurements and represents valuable data used for calibration of the flow model as it essentially dominates the flow directions. Consequently, transient effects in the head field can be of utmost importance. It is usually vertically averaged information, but in some cases can be known at different intervals along a vertical.

7. Location of the boundary of the flow domain being investigated: This information is obtained from a regional hydrogeological and hydrological investigation. The boundaries are often chosen in such a manner that a feasible formulation of the boundary conditions (fixed head or streamline) can be obtained. The location of the boundary of the aquifer is based on a regional hydrogeological and hydrological assessment and is always subject to some

uncertainty, particularly if one or more of the boundaries are a groundwater divide.

8. Boundary conditions – generally, the heads at the boundary (or portions of it) or of the water flux through the boundary (or portions of it): This information can be obtained from hydrological and hydrogeological investigations. Fixed head boundary conditions are subject to uncertainty caused by data interpretation and interpolation. The transient behaviour of these conditions cannot often be assessed in detail. Flux boundary conditions are also difficult to estimate. They can often be determined in a satisfactory manner with the help of flow models, provided that reliable data of hydraulic conductivity and piezometric head are available. Nevertheless, some uncertainty inevitably remains. The averaged value and the transient behaviour of both types of boundary conditions can be important for the flow field, and, therefore, for the location of the capture zone or catchment.

9. Hydraulic conductivity (or transmissivity) of the aquifer: This information can be obtained from pumping test evaluation or other procedures. Hydraulic conductivity always shows a more or less pronounced spatial variability due to the heterogeneous nature of aquifers. Therefore, a thorough investigation of field-scale hydraulic conductivity is advisable. Spatial variability can considerably affect the uncertainty of the location of the capture zone or catchment. In addition, the scale at which the measurements have been taken has to be carefully considered in the evaluation of the measurement.

10. Aquifer porosity: This information is relevant for proper prediction of isochrones and can be deduced, for instance, from tracer tests. Aquifer kinematic porosity directly affects the flow velocity and therefore the residence times, which subsequently determine the location of the capture zone. Moreover, spatial variability of field scale and local porosity may also exist. However, the effect of spatial variability of porosity tends to be smaller than that of hydraulic conductivity.

6.3.3 Representation of uncertainty

Many of the above items are obtained from localised information, such as that typically measured in boreholes. Therefore, the quality of information over an entire flow domain will very much depend on the spatial and/or temporal density of the available data. However, for economic and logistic reasons such information is often sparse. The location of data points is normally restricted to a limited set of (convenient) locations within the aquifer. Similarly, the temporal frequency of measurements is often limited. Moreover, experimental data are sometimes corrupted by measurement and interpretive errors. Overall, the combined effects of the uncertainty of all parameters and

conditions can considerably affect the precision of the calculated capture zone or catchment.

For small areas a more simplified and intuitive assessment of the uncertainty is often possible, which can be taken into account in the delineation of the protection zone. However, for larger areas the uncertainty can be quite large. Depending on the economical and ecological importance of the protection zone, the implications of the degree of uncertainty associated with its predicted location can be prohibitive. Therefore, methods are needed to quantify the uncertainty and provide guidance on the acquisition of site data to reduce it. In general, uncertainty can be reduced by increasing investigation. However, since resources are limited the task should be based on a pragmatic methodological approach, in the sense that it optimises efficiency. Consequently, there is a need for knowledge and tools that enable a conceptual and quantitative assessment of the impact of parameter uncertainty on the location of existing or planned protection zones.

Over the last decade there have been a number of important studies that have provided the required knowledge and tools, both analytical and numerical, for quantifying the effects of the above uncertainties on source protection zones; these have been summarised by Stauffer et al. (2005). Much of this work has focused on the effects of aquifer heterogeneity (van Leeuwen et al., 1998), where transmissivity and/or hydraulic conductivity is treated as a spatially correlated random variable. Various numerical or analytical Monte Carlo techniques have been proposed to deduce capture zones or catchments in a statistical manner, for example by Kunstmann and Kinzelbach (2000), Franzetti and Guadagnini (1996), Guadganini and Franzetti (1999), van Leeuwen et al. (1999), Wheater et al. (2000), Hunt et al. (2001), Feyen et al. (2001), Jacobson et al. (2002), and Feyen et al. (2003).

The consequence of the uncertainty of the essential parameters that determine capture zones or catchments of a pumping well is that the location of these zones cannot be determined with certainty. Therefore, the location of the protection zones can only be defined in a statistical manner. Although studies have shown that while with increasing amounts of data these uncertainties can be reduced, the amount of information required can be prohibitive (e.g. Bakr and Butler, 2004; van Leeuwen et al., 2000). It is recommended that the most suitable way of handling these uncertainties is to provide a probability map (Figure 6.4), rendering the probability with which a particular location belongs to the capture zone or catchment. Such concepts can then be fed directly into a risk assessment of a particular groundwater resource (Stauffer et al., 2005).

6.3.4 Arid and semi-arid zone applications

From the data requirements listed in Section 6.3.2, it is clear that there are major challenges in determining (and then implementing

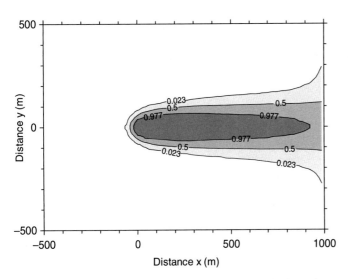

Figure 6.4. Probability map of a well catchment; Monte Carlo results (well located at $x = 0$, $y = 0$) (from Stauffer *et al.*, 2005).

management decisions associated with) well protection zones in arid and semi-arid regions. This has frequently meant that determination of protection zones for specific groundwater sources has been neglected altogether. This is unfortunate as most groundwater supplies are from shallow wells or springs and therefore highly vulnerable to the effects of surface contamination (ARGOSS, 2001). In addition, as water drawn from these supplies often receives minimal treatment, the risk of faecal–oral transmission can be high. Therefore, where well protection has been considered it has either focused on well construction (Lewis and Chilton, 1984) or on simple rules that focus on protection against pathogenic organisms arising from sewage (Taylor, 2005). For example, Lewis *et al.* (1982) suggest minimum distances of between 10 and 15 m between pit latrines and well or spring in order reduce the risk of transmission of microbial pathogens and Robins *et al.* (2007) propose a similar figure of 10 m based on work undertaken in Uganda. However, with increasing demand on water resources and a need for long-term security of groundwater supplies, more sophisticated approaches to groundwater protection which integrate groundwater vulnerability with source protection in a consistent manner are required. Thus, there is a need for relatively simple modelling approaches that can provide estimates of contaminant travel time both through the unsaturated zone and through the underlying groundwater system in order to assess contaminant travel times of pollutants affecting groundwater as a whole and groundwater supplies in particular.

6.4 SUMMARY

Groundwater is a vital resource in arid and semi-arid regions and will continue to be so for the foreseeable future due to increased demands for water resources as a result of rising population numbers and living standards. Consequently, ensuring the long-term security of aquifers in these areas is essential, and this includes the maintenance of good water quality. This requires protection of underlying groundwater from contamination arising from surface activities.

To do this in a cost-effective and targeted manner requires the identification of those groundwater bodies that are important as a resource and at risk from surface pollution, together with a method for ranking these aspects. Groundwater vulnerability mapping provides a distributed quantitative or qualitative measure of aquifer presence and risk (generic or specific) of contamination, allowing areas on the surface to be identified where measures can be prioritised in order to protect groundwater resources.

Commensurate with these is the determination of well capture and catchment zones, which allow the protection of existing groundwater sources at a more localised level. Unfortunately, the implementation of these approaches in arid and semi-arid environments is particularly challenging on two counts. The first is that frequently the data requirements to support these measures are insufficient. This may be due to poor data collection, but it is exacerbated by the extreme and heterogeneous nature of the geological environment and the hydrological processes that operate within this environment. Although there are opportunities from increasing availability of remotely sensed data, insufficient information is likely to remain a challenge. Consequently, there is a continuing need for the refinement of suitable techniques for use in arid environments and the development of suitable tools to help quantify uncertainties associated with this lack of data (NRC, 1993).

The other challenge lies in the will of decision makers to fund vulnerability studies and then utilise the results in their decision making. Allied with this is the need to communicate the importance of protection measures to those living and working in vulnerable areas. Although this still presents many great challenges, there are indications that this is increasingly being taken seriously, with a growing number of vulnerability studies being reported and the targeting of specific problems (e.g. nitrates and organic pollutants) being undertaken. It is hoped that such endeavours will continue and lead to the long-term benefit of those who live and work in such areas.

REFERENCES

Adams, B. and Foster, S. S. D. (1992) Land-surface zoning for groundwater protection. *Journal of the Institute of Water and Environmental Management* **6**, 312–320.

Al-Adamat, R. A. N., Foster, I. D. L. and Baban, S. M. J. (2003) Groundwater vulnerability and risk mapping for the Basaltic aquifer of the Azraq basin of Jordan using GIS, Remote sensing and DRASTIC. *Applied Geography* **23**, 303–324.

Al-Hanbali, A. and Kondoh, A. (2008) Groundwater vulnerability assessment and evaluation of human activity impact (HAI) within the Dead Sea groundwater basin, Jordan. *Hydrogeology Journal* 16, 499–510.

Al Kuisi, M., El-Naqa, A. and Hammouri, N. (2006) Vulnerability mapping of shallow groundwater aquifer using SINTACS model in the Jordan Valley area, Jordan. *Environmental Geology* 50, 651–667.

Al-Zabet, T. (2002) Evaluation of aquifer vulnerability to contamination potential using the DRASTIC method. *Environmental Geology* 43, 203–208.

Alemaw, B. F., Shemang, E. M. and Chaoka, T. R. (2004) Assessment of groundwater pollution vulnerability and modelling of the Kanye Wellfield in SE Botswana – a GIS approach. *Physics and Chemistry of the Earth, Parts A/B/C* 29, 1125–1128.

Aller, L., Bennett, T., Lehr, J. H., Petty, R. J. and Hackett, G. (1987) *DRASTIC: A Standardised System for Evaluating Groundwater Pollution Potential Using Hydrogeological Settings.* US-EPA Report.

Almasri, M. N. (2008) Assessment of intrinsic vulnerability to contamination for Gaza coastal aquifer, Palestine. *Journal of Environmental Management* 88, 577–593.

Andersen, L. J. and Gosk, E. (1989) Applicability of vulnerability maps. *Environmental Geology* 13, 39–43.

ARGOSS (2001) *Guidelines for Assessing the Risk to Groundwater from On-Site Sanitation.* British Geological Survey.

Arnold, J. G., Allen, P. M., Muttiah, R. and Bernhardt, G. (1995) Automated base flow separation and recession analysis techniques. *Ground Water* 33, 1010–1018.

Bakr, M. I. and Butler, A. P. (2004) Worth of head data in well-capture zone design: deterministic and stochastic analysis. *Journal of Hydrology* 290, 202–216.

Barber, C., Bates, L. E., Barron, R. and Allison, H. (1993) Assessment of the relative vulnerability of groundwater to pollution: a review. *Journal of Australian Geology and Geophysics* 14, 1147–1154.

Bear, J. and Jacobs, M. (1965) On the movement of water bodies injected into aquifers. *Journal of Hydrology* 3, 37–57.

Brosig, K., Geyer, T., Subah, A. and Sauter, M. (2008) Travel time based approach for the assessment of vulnerability of karst groundwater: the transit time method. *Environmental Geology* 54, 905–911.

Chave, P., Howard, G., Schijven, J. *et al.* (2006) Groundwater protection zones. In *Protecting Groundwater for Health: Managing the Quality of Drinking-water Sources*, ed. O. Schmoll, G. Howard, J. Chilton and I. Chorus, 465–492. World Health Organization.

Chilton, J. (2006) Assessment of aquifer pollution vulnerability and susceptibility to the impacts of abstraction. In *Protecting Groundwater for Health: Managing the Quality of Drinking-water Sources*, ed. O. Schmoll, G. Howard, J. Chilton and I. Chorus, 199–241. World Health Organization.

Chilton, P. J. and Foster, S. S. D. (1995) Hydrogeological characterisation and water-supply potential of basement aquifers in tropical Africa. *Hydrogeology Journal* 3, 36–49.

Chitsazan, M. and Akhtari, Y. (2009) A GIS-based DRASTIC model for assessing aquifer vulnerability in Kherran plain, Khuzestan, Iran. *Water Resources Management* 23, 1137–1155.

Civita, M. and de Maio, M. (2000) *SINTACS R5, A New Parametric System for the Assessment and Automating Mapping of Groundwater Vulnerability to Contamination.* Pitagora Editor (Bologna).

Dawoud, M. A. (2004) Design of national groundwater quality monitoring network in Egypt. *Environmental Monitoring and Assessment* 96, 99–118.

Debernardi, L., De Luca, D. A. and Lasagna, M. (2008) Correlation between nitrate concentration in groundwater and parameters affecting aquifer intrinsic vulnerability. *Environmental Geology* 55, 539–558.

Denny, S., Allen, D. and Journeay, J. (2007) DRASTIC-Fm: a modified vulnerability mapping method for structurally controlled aquifers in the southern Gulf Islands, British Columbia, Canada. *Hydrogeology Journal* 15, 483–493.

Dimitriou, E. and Zacharias, L. (2006) Groundwater vulnerability and risk mapping in a geologically complex area by using stable isotopes, remote sensing and GIS techniques. *Environmental Geology* 51, 309–323.

Dixon, B. (2005) Groundwater vulnerability mapping: a GIS and fuzzy rule based integrated tool. *Applied Geography* 25, 327–347.

Doerfliger, N., Jeannin, P. Y. and Zwahlen, F. (1999) Water vulnerability assessment in karst environments: a new method of defining protection areas using a multi-attribute approach and GIS tools (EPIK method). *Environmental Geology* 39, 165–176.

EU (2009) http://ec.europa.eu/environment/water/water-framework/groundwater.html.

Evans, B. M. and Myers, W. L. (1990) A GIS-based approach to evaluating regional groundwater pollution potential with DRASTIC. *Journal of Soil and Water Conservation* 45, 242–245.

Evers, S. and Lerner, D. N. (1998) How uncertain is our estimate of a wellhead protection zone? *Ground Water* 36, 49–57.

Feyen, L., Beven, K. J., De Smedt, F. and Freer, J. (2001) Stochastic capture zone delineation within the generalized likelihood uncertainty estimation methodology: conditioning on head observations. *Water Resources Research* 37, 625–638.

Feyen, L., Ribeiro, P. J., Gómez-Hernández, J. J., Beven, K. J. and De Smedt, F. (2003) Bayesian methodology for stochastic capture zone delineation incorporating transmissivity measurements and hydraulic head observations. *Journal of Hydrology* 271, 156–170.

Foster, S. S. D. (1987) Fundamental concepts in aquifer vulnerability pollution risk and protection strategy. In *Vulnerability of Soils and Groundwater to Pollution*, ed. W. van Duijvenbooden and H. G. van Waegeningh, 38, 69–86. TNO Committee on Hydrological Research, The Hague.

Foster, S. S. D. and Hirata, R. (1998) *Groundwater Pollution Risk Assessment.* Pan American Centre for Sanitary Engineering and Environmental Sciences, Lima.

Foster, S. S. D. and Skinner, A., C. (1995) Groundwater protection: the science and practice of land surface zoning. In *Groundwater Quality: Remediation and Protection*, IAHS Publ. 225, 471–482.

Franzetti, S. and Guadagnini, A. (1996) Probabilistic estimation of well catchments in heterogeneous aquifers. *Journal of Hydrology* 174, 149–171.

Fredrick, K. C., Becker, M. W., Flewelling, D. M., Silavisesrith, W. and Hart, E. R. (2004) Enhancement of aquifer vulnerability indexing using the analytic-element method. *Environmental Geology* 45, 1054–1061.

Fritch, T. G., McKnight, C. L., Yelderman Jr, J. C. and Arnold, J. G. (2000) An aquifer vulnerability assessment of the Paluxy aquifer, central Texas, USA, using GIS and a modified DRASTIC approach. *Environmental Management* 25, 337–345.

Gogu, R. C. and Dassargues, A. (2000) Current trends and future challenges in groundwater vulnerability assessment using overlay and index methods. *Environmental Geology* 39, 549–559.

Goldscheider, N., Klute, M., Sturm, S. and Hötzl, H. (2000) The PI method: a GIS based approach to mapping groundwater vulnerability with special consideration of karst aquifers. *Z Angew Geol.* 463, 157–166.

Guadagnini, A. and Franzetti, S. (1999) Time-related capture zones for contaminants in randomly heterogeneous formations. *Ground Water* 37, 253–260.

Hamza, M. H., Added, A., Rodríguez, R., Abdeljaoued, S. and Ben Mammou, A. (2007) A GIS-based DRASTIC vulnerability and net recharge reassessment in an aquifer of a semi-arid region (Metline-Ras Jebel-Raf Raf aquifer, northern Tunisia). *Journal of Environmental Management* 84, 12–19.

Hiscock, K. M., Lovett, A. A., Brainard, J. S. and Parahat, J. P. (1995) Groundwater vulnerability assessment: two case studies using GIS methodology. *Quarterly Journal of Engineering Geology* 28, 179–194.

Hölting, B., Haertle, T., Hohberger, K. H., Nachtigall Weinzierl, W. and Wrobel, J. P. (1995) Konzept zur Schutzfunktion der Grundwasserüberdeckung [Concept for the evaluation of the protective function of the unsaturated above the water table]. *Geol Jahrb* C63, 5–24.

Hunt, R. J., Steuer, J. J., Mansor, M. T. C. and Bullen, T. D. (2001) Delineating a recharge area for a spring using numerical modeling, Monte Carlo techniques, and geochemical investigation. *Ground Water* 39, 702–712.

Jacobson, E. andricevic, R. and Morrice, J. (2002) Probabilistic capture zone delineation based on an analytic solution. *Ground Water* 40, 85–95.

Jamrah, A., Al-Futaisi, A., Rajmohan, N. and Al-Yaroubi, S. (2008) Assessment of groundwater vulnerability in the coastal region of Oman using DRASTIC index method in GIS environment. *Environmental Monitoring and Assessment* 147, 125–138.

Kinzelbach, W., Marburger, M. and Chiang, W.-H. (1992) Determination of groundwater catchment areas in two and three spatial dimensions. *Journal of Hydrology* 134, 221–246.

Kunstmann, H. and Kinzelbach, W. (2000) Computation of stochastic wellhead protection zones by combining the first-order second-moment method and Kolmogorov backward equation analysis. *Journal of Hydrology* 237, 127–146.

Lewis, W. J. and Chilton, P. J. (1984) Performance of sanitary completion measures of wells and boreholes used for rural water supplies in Malawi. In *Challenges in African Hydrology and Water Resources*, ed. D. E. Walling, S. S. D. Foster and P. Wurzel, 144, 235–247. IAHS, Harare.

Lewis, W. J., Foster, S. S. D. and Draser, B. S. (1982) *The Risk of Ground-water Pollution by On-Site Sanitation in Developing Countries*. International Reference Center for Waste Disposal (IRCWD), Dübendorf, Switzerland.

Margane, A., Hobler, M. and Subah, A. (1999) Mapping of groundwater vulnerability and hazards to groundwater in the Irbid Area, N. Jordan. *Zeitschrift für Angewandte Geologie* **45**, 175–187.

McLay, C. D. A., Dragten, R., Sparling, G. and Selvarajah, N. (2001) Predicting groundwater nitrate concentrations in a region of mixed agricultural land use: a comparison of three approaches. *Environmental Pollution* **115**, 191–204.

Merchant, J. W. (1994) GIS-based groundwater pollution hazard assessment: a critical review of the DRASTIC model. *Photogrammetric Engineering and Remote Sensing* **60**, 1117–1127.

Murray, K. E. and McCray, J. E. (2005) Development and application of a regional-scale pesticide transport and groundwater vulnerability model. *Environmental and Engineering Geoscience* **11**, 271–284.

Neukum, C., Hötzl, H. and Himmelsbach, T. (2008) Validation of vulnerability mapping methods by field investigations and numerical modelling. *Hydrogeology Journal* **16**, 641–658.

NRA (1992) *Policy and Practice in the Protection of Groundwater*. NRA, Bristol, UK.

NRC (1993) *Groundwater Vulnerability Assessment: Contamination Potential under Conditions of Uncertainty*. National Academy Press.

Panagopoulos, G., Antonakos, A. and Lambrakis, N. (2006) Optimization of the DRASTIC method for groundwater vulnerability assessment via the use of simple statistical methods and GIS. *Hydrogeology Journal* **14**, 894–911.

Piscopo, G. (2001) *Groundwater Vulnerability Map, Explanatory Notes, Castlereagh Catchment, NSW*. Rep. No. CNR 2001.017. Centre for Natural Resources, Department of Land and Water Conservation, Australia.

Reddy, K. R. (2008) Physical and Chemical Groundwater Remediation Technologies. In *Overexploitation and Contamination of Shared Groundwater Resources*, 257–274.

Robins, N. S. (1998) Recharge: the key to groundwater pollution and aquifer vulnerability. In *Groundwater Pollution, Aquifer Recharge and Vulnerability*, ed. N. S. Robins, 1–5. Geological Society Special Publications.

Robins, N. S., Chilton, P. J. and Cobbing, J. E. (2007) Adapting existing experience with aquifer vulnerability and groundwater protection for Africa. *Journal of African Earth Sciences* **47**, 30–38.

Rupert, M. G. (2001) Calibration of the DRASTIC ground water vulnerability mapping method. *Ground Water* **39**, 625–630.

Secunda, S., Collin, M. L. and Melloul, A. J. (1998) Groundwater vulnerability assessment using a composite model combining DRASTIC with extensive agricultural land use in Israel's Sharon region. *Journal of Environmental Management* **54**, 39–57.

Simmers, I. (1998) *Groundwater Recharge: An Overview of Estimation Problems and Recent Developments*. Geological Society, London, Special Publications **130**, 107–115.

Stauffer, F., Guadagnini, A., Butler, A. *et al.* (2005) Delineation of source protection zones using statistical methods. *Water Resources Management* **19**, 163–185.

Stigter, T. (2006) Evaluation of an intrinsic and a specific vulnerability assessment method in comparison with groundwater salinisation and nitrate contamination levels in two agricultural regions in the south of Portugal. *Hydrogeology Journal* **14**, 79–99.

Taylor, R. (2005) Groundwater protection in sub-Saharan Africa. *Waterlines* **24**, 21–23.

van Leeuwen, M., te Stroet, C. B. M., Butler, A. P. and Tompkins, J. A. (1998) Stochastic determination of well capture zones. *Water Resources Research* **34**, 2215–2223.

van Leeuwen, M., te Stroet, C. B. M., Butler, A. P. and Tompkins, J. A. (1999) Stochastic determination of the Wierden (Netherlands) capture zones. *Ground Water* **37**, 8–17.

van Leeuwen, M., Butler, A. P., te Stroet, C. B. M. and Tompkins, J. A. (2000) Stochastic determination of well capture zones conditioned on regular grids of transmissivity measurements. *Water Resources Research* **36**, 949–957.

van Stemport, D., Ewert, L. and Wassenaar, L. (1993) Aquifer vulnerability index. A GIS-compatible method for groundwater vulnerability mapping. *Canadian Water Resources Journal* **18**, 25–37.

Vías, J. M., Andreo, B., Perles, M. J. and Carrasco, F. (2005) A comparative study of four schemes for groundwater vulnerability mapping in a diffuse flow carbonate aquifer under Mediterranean climatic conditions. *Environmental Geology* **47**, 586–595.

Vrba, J. and Zaporozec, A. (eds.) (1994) *Guidebook on Mapping Groundwater Vulnerability*, **16**, 1–131. International Association of Hydrogeologists.

Wang, Y., Merkel, B., Li, Y. *et al.* (2007) Vulnerability of groundwater in Quaternary aquifers to organic contaminants: a case study in Wuhan City, China. *Environmental Geology* **53**, 479–484.

Werz, H. and Hötzl, H. (2007) Groundwater risk intensity mapping in semi-arid regions using optical remote sensing data as an additional tool. *Hydrogeology Journal* **15**, 1031–1049.

Wheater, H. S., Tompkins, J. A., van Leeuwen, M. and Butler, A. P. (2000) Uncertainty in groundwater flow and transport modelling – a stochastic analysis of well-protection zones. *Hydrological Processes* **14**, 2019–2029.

Williams, A., Bloomfield, J., Griffiths, K. and Butler, A. (2006) Characterising the vertical variations in hydraulic conductivity within the Chalk aquifer. *Journal of Hydrology* **330**, 53–62.

Xu, Y. and Braune, E. (1995) *A Guideline for Groundwater Protection for the Community Water Supply and Sanitation Programme*. Pretoria.

7 Variable density groundwater flow: from modelling to applications

C. T. Simmons, P. Bauer-Gottwein, T. Graf, W. Kinzelbach, H. Kooi, L. Li, V. Post, H. Prommer, R. Therrien, C. I. Voss, J. Ward and A. Werner

7.1 INTRODUCTION

Arid and semi-arid climates are mainly characterised as those areas where precipitation is less (and often considerably less) than potential evapotranspiration. These climate regions are ideal environments for salt to accumulate in natural soil and groundwater settings since evaporation and transpiration essentially remove freshwater from the system, leaving residual salts behind. Similarly, the characteristically low precipitation rates reduce the potential for salt to be diluted by rainfall. Thus arid and semi-arid regions make ideal 'salt concentrator' hydrologic environments. Indeed, salt flats, playas, sabkhas and saline lakes, for example, are ubiquitous features of arid and semi-arid regions throughout the world (Yechieli and Wood, 2002). In such settings, variable density flow phenomena are expected to be important, especially where hypersaline brines overlie less dense groundwater at depth. In contrast, seawater intrusion in coastal aquifers is a global phenomenon that is not constrained to only arid and semi-arid regions of the globe and is inherently a variable density flow problem by its very nature. These two examples make it clear that variable density flow problems occur in, but importantly extend beyond, arid and semi-arid regions of the globe. The intention of this chapter is therefore not to limit ourselves to modelling arid zone hydrological systems, but rather to present a more general treatment of variable density groundwater flow and solute transport phenomena and modelling. The concepts presented in this chapter are therefore not climatologically constrained to arid or semi-arid zones of the world, although they do apply equally there. The objective of the chapter is to illustrate the state of the field of variable density groundwater flow as it applies to both current modelling and applications. Throughout, we highlight approaches and current resolutions, and explicitly point to future challenges in this area.

7.2 IMPORTANCE OF VARIABLE DENSITY FLOW

It has long been known that the density of a fluid can be modified by changing its solute concentration, temperature and, to a lesser extent, its pressure. Indeed, researchers in the area of fluid mechanics have studied how fluid density affects flow behaviour for well over a century, both in fluids only and in fluid-filled porous media. It was understood that areas of application for these studies was wide and varied including applications to astrophysics, chemical reactor engineering, energy storage and recovery, geophysics, geothermal reservoirs, material science, metallurgy, nuclear waste disposal and oceanography, to name just a few.

More recently, there has been a massive explosion in the field because of worldwide concern about the future of our energy resources and the pollution of our environment. These have been a principal catalyst for the application of earlier developments made in traditional fluid mechanics to the study of groundwater hydrology. In the preface to their popular text *Convection in Porous Media*, Nield and Bejan (1999) state that 'Papers on convection in porous media continue to be published at a rate of over 100 per year'. Table 7.1 summarises some key areas in variable density flow research in groundwater hydrology in particular. The listed references are intended to be illustrative rather than exhaustive. This table clearly demonstrates the widespread importance, diversity, extent and interest in applications of variable density flow phenomena in groundwater hydrology. As noted in Simmons (2005), key areas include seawater intrusion, fresh–saline water interfaces and saltwater upconing in coastal aquifers, subterranean groundwater discharge, dense contaminant plume migration, DNAPL studies, density-driven transport in the vadose zone, flow through salt formations in high-level radioactive waste disposal

Groundwater Modelling in Arid and Semi-Arid Areas, ed. Howard S. Wheater, Simon A. Mathias and Xin Li. Published by Cambridge University Press.

Table 7.1. *Some key physical systems where variable density groundwater flow phenomena are important in groundwater hydrology.*

Variable density groundwater system	Relevant papers
Sea water intrusion, fresh–saline water interfaces in coastal aquifers	Yechieli *et al.* (2001)
	Kooi *et al.* (2000)
	Post and Kooi (2003)
	Underwood *et al.* (1992)
	Voss and Souza (1987)
	Huyakorn *et al.* (1987)
	Pinder and Cooper (1970)
	Werner and Gallagher (2006)
Subterranean groundwater discharge	Langevin (2003)
	Kaleris *et al.* (2002)
Infiltration of leachates from waste disposal sites, dense contaminant plumes	Liu and Dane (1996)
	Zhang and Schwartz (1995)
	Oostrom *et al.* (1992a,b)
	Koch and Zhang (1992)
	Schincariol and Schwartz (1990)
	Pashcke and Hoopes (1984)
	Le Blanc (1984)
	Frind (1982a)
DNAPL flow and transport	Li and Schwartz (2004)
	Lemke *et al.* (2004)
	Oostrom *et al.* (2003)
Density-driven transport in the vadose zone	Ying and Zheng (1999)
	Ouyang and Zheng (1999)
Flow through salt formations in high-level disposal sites, heat and solute movement near salt domes	Jackson and Watson (2001)
	Williams and Ranganathan (1994)
	Hassanizadeh and Leijnse (1988)
Heat and fluid flow in geothermal systems	Oldenburg and Pruess (1999)
	Gvirtzman *et al.* (1997)
Sedimentary basin mass and heat transport processes, diagenesis processes	Garven *et al.* (2003)
	Sharp *et al.* (2001)
	Raffensperger and Vlassopoulous (1999)
	Wood and Hewett (1984)
Palaeohydrogeology of sedimentary basins	Senger (1993)
	Gupta and Bair (1997)
Processes beneath playas, sabkhas and playa lakes	Yechieli and Wood (2002)
	Sanford and Wood (2001)
	Simmons *et al.* (1999)
	Wooding *et al.* (1997a,b)
	Duffy and Al-Hassan (1988)
Operation of saline (and irrigation) water disposal basins	Simmons *et al.* (2002)
Density affects in applied tracer tests	Barth *et al.* (2001)
	Zhang *et al.* (1998)
	Istok and Humphrey (1995)
	Le Blanc *et al.* (1991)

sites, heat and fluid flow in geothermal systems, palaeohydrogeology of sedimentary basins, sedimentary basin mass and heat transport and diagenesis, processes beneath sabkhas and salt lakes and buoyant plume effects in applied tracer tests. The

modelling of these physical systems is central to hydrogeologic analyses.

Simmons (2005) presented an overview of variable density flow phenomena, including current challenges, and future possibilities for this field of groundwater investigation. The widespread importance and interest in variable density flow phenomena in groundwater hydrology are also illustrated in exhaustive review articles on this topic by Diersch and Kolditz (2002) and Simmons *et al.* (2001). In addition, the text by Holzbecher (1998) is an excellent source of material relating to the modelling of density dependent flows in porous media. The reader is referred to these texts for an exhaustive review of this subject matter. In particular, Diersch and Kolditz (2002) recently reviewed the state of the art in modelling of density-driven flow and transport in porous media. They discuss conceptual models for density-driven convection systems, governing balance equations, critical phenomenological laws and constitutive relations for fluid density and viscosity, and the numerical methods employed for solving the resulting non-linear problems. Furthermore, these authors discuss the major limitations of current model-verification (benchmarking) test cases used in the testing of numerical models and methods. Diersch and Kolditz (2002) and Simmons (2005) provide detailed treatments of the current challenges that face the field of variable density groundwater flow phenomena, from both physical and modelling viewpoints. The reader is referred to those discussions as an accompaniment to this text.

7.3 A BRIEF HISTORICAL PERSPECTIVE

Nield and Bejan (2006) provide an excellent discussion of the topic of convection in porous media including historical and new conceptual developments within this field. Some of the earliest work was carried out in the field of traditional fluid mechanics. A logical trend emerged in the way this field evolved. Earliest work in the early 1900s involved heated fluids only (Benard, 1900; Rayleigh, 1916), followed by a move to combined heat and porous media studies in the 1940s (Horton and Rogers, 1945; Lapwood, 1948), solute and porous media studies in the 1950s and 1960s (Wooding, 1959, 1962, 1963, 1969) and finally combined thermohaline studies in porous media in the 1960s (Nield, 1968).

Most of these studies were designed to discover the most fundamental aspects of the behaviour of variable density flow behaviour and were typically either laboratory or analytically based. In comparison, as groundwater hydrologists, our interest in energy and solute transport is relatively recent (post 1950s in general). Therefore, it is challenging to find much evidence of interest and exploration in variable density flow groundwater applications at all throughout the earlier historical development of groundwater hydrology. But there is one clear exception – the problem of seawater intrusion.

Some of the earliest pioneering work in this area traces back to that of Badon-Ghyben (1888) and Herzberg (1901), whose analyses could be used to determine the steady position of saltwater–freshwater interfaces in coastal aquifers. Some of the earliest analytical solutions for the sharp saltwater–freshwater interface in an infinitely thick confined aquifer emerged in the 1950s (Glover, 1959; Henry, 1959). In the problem of seawater intrusion, Hele-Shaw cell analogues (Bear, 1972) were used in some of the earliest exploratory studies of seawater intrusion and clearly predated numerical solutions to these problems. Some of the earliest numerical analyses emerged in studies by Pinder and Cooper (1970), Segol and Pinder (1976), Huyakorn and Taylor (1976) and Frind (1982a, 1982b).

Table 7.1 illustrates that the interest in variable density groundwater flow applications has grown to include many other important areas of hydrogeology. In more recent times, groundwater modelling codes have begun to employ a range of new and improved numerical techniques and this coupled to faster computers with larger memories has allowed for the simulation and solution of more complex problems, with fully coupled flow and solute transport in even 3D cases. Recent developments in modelling efforts have included simultaneous heat and transport in thermohaline convection problems (Diersch and Kolditz, 2002; Graf and Therrien, 2007b), adaptation of these to heterogeneous systems including fractured rock hydrology (Graf and Therrien, 2005; Graf, 2005) and even chemically reactive transport (Freedman and Ibaraki, 2002; Post and Prommer, 2007). Indeed, it may be argued that one major limitation of the current field of variable density flow is not necessarily the availability of numerical models with the inherent capacity to solve these sorts of very complex problems, but is perhaps the availability of real data to occupy, test and verify the emerging new generation of computer models.

7.4 VARIABLE DENSITY FLOW PHYSICS

7.4.1 Overview

Field, laboratory and modelling studies demonstrate that fluid density gradients caused by variations in concentration and temperature (and to a much lesser degree fluid pressure) play a significant role in solute and/or heat transport in groundwater systems. Figures 7.1 and 7.2 show how fluid density varies as a function of concentration and temperature.

It is easy to accept that fluid density should be at least one factor influencing groundwater flow processes, because in many systems groundwater concentrations can vary by several orders of magnitude (e.g. salt lakes, seawater intrusion), and in all cases

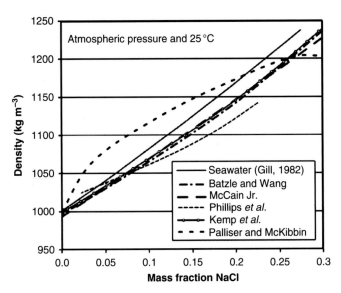

Figure 7.1. Brine density as a function of NaCl mass fraction, computed for the conditions of atmospheric pressure at temperature of 25 °C. Source: Adams and Bachu (2002).

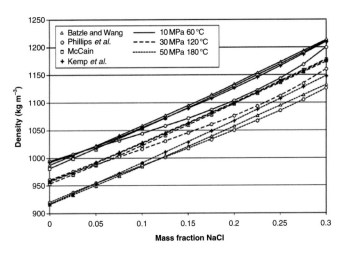

Figure 7.2. Brine density calculated at conditions characteristic of some typical sedimentary basins. Source: Adams and Bachu (2002).

the Earth's geothermal gradient has the potential to set up circulations where warm fluid at depth can rise and shallower cooler fluids can sink. Simmons (2005) considered a typical groundwater hydraulic gradient of, say, a 1 m hydraulic head drop over a lateral distance of a kilometre (i.e. a gradient of 1 in 1000) and showed that the equivalent 'driving force' in density terms would be a density difference of 1 kg m^{-3} relative to a reference density of freshwater whose density is 1000 kg m^{-3}. This is a solution whose concentration is only about 2 g l^{-1} (about 5% of seawater!), which is quite dilute in comparison to many plumes one may encounter in groundwater studies.

7.4.2 Possible manifestations of variable density flow

Whether density-driven flow is important in a physical hydrogeologic setting is determined by a number of competing factors. These involve a complex interplay between fluid properties, the interaction within mixed convective flow of fluid motion driven by both density differences (free convection) and hydraulic gradients (advection or forced convection) and, finally, the properties of the porous medium. The ability to maintain a density gradient and a convective flow regime is dissipated by dispersive mixing processes. The density contrast and its stable or unstable configuration (e.g. light fluid overlying dense fluid as in a stable case of seawater intrusion, or dense fluid overlying light fluid as in the potentially unstable case of a leachate plume migrating from a landfill) is also of critical importance. These will directly affect solute transport processes and the way in which variable density flow processes are subsequently manifested.

For example, in a steady-state coastal setting, density effects create a seawater wedge that penetrates inland for some distance. Although tidal effects do create mixing, a thin interfacial mixing-layer is maintained by a slow flow of less saline groundwater over the wedge. This sort of rotational flow associated with seawater intrusion phenomena is the manifestation of variable density flow physics and may be referred to as the 'Henry circulation' (Henry, 1964).

In some specialised cases, where there are large unstable density inversions, transport can be characterised by rapid instability development where lobe-shaped instabilities (fingers) form and sink under gravitational influence resulting in enhanced solute transport and greater mixing compared to diffusion alone (Simmons *et al.*, 2001).

The complexity of the problems involved generally increases as one moves from situations where the light fluid overlies the dense fluid to potentially unstable situations where the dense fluid overlies the less dense fluid. The challenges for numerical simulation appear to grow in the same way, as will be discussed in later sections of this chapter.

7.4.3 Dimensionless numbers

It is important to note that, in all cases, the propensity for density-driven flow is largest where there are large density gradients and the porous media is of high permeability. In contrast, dispersion acts to reduce the density gradient and therefore to dissipate density-driven flow. Textbooks such as those by Nield and Bejan (2006) and Holzbecher (1998) provide a comprehensive overview of the theoretical treatment of density-driven flow in porous media. Critical dimensionless numbers such as the Rayleigh number (*Ra*) and mixed convection ratio (*M*) are important parameters used in the characterisation of such systems.

In simple free convective systems, where mechanical dispersion is assumed independent of convective flow velocity, the onset of instability is determined by the value of a non-dimensional number called the Rayleigh number (Ra). This dimensionless number is the ratio between buoyancy-driven forces and resistive forces caused by diffusion and dispersion.

Ra is defined by (e.g. Simmons et al., 2001):

$$Ra = \frac{U_c H}{D_0} = \frac{gk\beta(C_{max} - C_{min})H}{\theta \nu_0 D_0} = \frac{\text{Buoyancy and gravitation}}{\text{Diffusion and dispersion}} \quad (7.1)$$

where U_c is the convective velocity, H is the thickness of the porous layer, D_0 is the molecular diffusivity, g is the acceleration due to gravity, k is the intrinsic permeability, $\beta = \rho_0^{-1}(\partial\rho/\partial C)$ is the linear expansion coefficient of fluid density with changing fluid concentration, C_{max} and C_{min} are the maximum and minimum values of concentration respectively, θ is the aquifer porosity and $\nu_0 = \mu_0/\rho_0$ is the kinematic viscosity of the fluid. Simmons et al. (2001) and Simmons (2005) discuss some of the difficulties encountered in the application of the Rayleigh number to practical field-based settings. For sufficiently high Rayleigh numbers greater than some critical Rayleigh number Ra_c, gravity induced instability will occur in the form of waves in the boundary layer that develop into fingers or plumes. This critical Rayleigh number defines the transition between dispersive/diffusive solute transport (at lower than critical Rayleigh numbers) and convective transport by density-driven fingers (at higher than critical Rayleigh numbers).

In many cases, the system we are studying is a mixed convective system where both free and forced convection operate to control solute distributions. In those cases, it is interesting to examine the relative strengths of each process in controlling the resultant transport process. The ratio of the density-driven convective flow speed to advective flow speed determines the dominant transport mechanism. In the case of an isotropic porous medium the mixed convection ratio may be written as (see, for example, Bear, 1972):

$$M = \frac{\left(\frac{\Delta\rho}{\rho_0}\right)}{\left(\frac{\Delta h}{\Delta L}\right)} = \frac{\text{Free convective speed}}{\text{Forced convective speed}} \quad (7.2)$$

If $M \gg 1$, then free convection is dominant. If $M \ll 1$, then forced convection is dominant. Where $M \sim 1$ they are of comparable magnitude. Here, $\Delta\rho$ is the density difference, ρ_0 is the lower reference density (both free convection parameters) and Δh is the hydraulic head difference measured over a length ΔL (both forced convection parameters).

7.4.4 The use of equivalent freshwater head in variable density analyses

Despite being one of the seemingly simplest methods of analysing variable density flow systems, the correct application of the equivalent freshwater head (EFH) is actually more complicated than is often understood. Post et al. (2007) present a detailed analysis on the use of hydraulic head measurements in variable density flow analyses. By way of introduction, freshwater head at point i may be defined as

$$h_{f,i} = z_i + \frac{P_i}{\rho_f g} \quad (7.3)$$

where z_i (elevation head) represents the (mean) level of the well screen, P_i is pressure of the groundwater at the well screen, ρ_f is freshwater density and g is acceleration due to gravity.

A common error that occurs in practice is for equivalent freshwater head to be used in the standard non-density-dependent version of Darcy's law that does not include a buoyancy gradient term critical in vertical flow calculations. This leads to errors in predicting flow rates and directions where vertical flow effects are of interest in vertically stratified layers (with respect to density) of fluid. In a recent issue paper by Post et al. (2007), it is noted that the use of EFH without care can be particularly problematic. Indeed, Lusczynski (1961), who originally formulated the concept, recognised that equivalent freshwater heads cannot be used to determine the vertical hydraulic gradient in an aquifer with water of non-uniform density. The term environmental water head (EWH) was established for this purpose (Lusczynski, 1961). However, it can be demonstrated that this environmental water head is equivalent to using equivalent freshwater head where the appropriate density-dependent version of Darcy's law is used (as was actually shown in Appendix 3 of the original Lusczynski study).

Post et al. (2007) argue that, provided that the proper corrections are taken into account, freshwater heads can be used to analyse both horizontal and vertical flow components. To avoid potential confusion, they recommended that the use of the so-called environmental water head, which was initially introduced to facilitate the analysis of vertical groundwater flow, be abandoned in favour of properly computed freshwater head analyses.

Indeed, some variable density groundwater flow simulators such as FEFLOW and SEAWAT (see Table 7.2) employ the equivalent freshwater head formulation in the variable density version of Darcy's law in the familiar form (for the vertical flow component):

$$q_z = -K_f \left[\frac{\partial h_f}{\partial z} + \left(\frac{\rho - \rho_f}{\rho_f} \right) \right] \quad (7.4)$$

where K_f is the freshwater hydraulic conductivity, h_f is the freshwater head, ρ is fluid density and ρ_f is freshwater density. Here the term $\frac{\rho - \rho_f}{\rho_f}$ represents the relative density contrast, and accounts for the buoyancy effect on the vertical flow. In the case of horizontal flow components, this term is neglected in the respective horizontal flow velocity formulations in equation (7.4).

Table 7.2. *Some numerical simulators for variable density groundwater flow and solute transport problems. This list is illustrative not exhaustive.*

Simulator	References for code development and verification
CFEST	Gupta *et al.* (1987)
FAST-C	Holzbecher (1998)
FEFLOW	Diersch (1988); Diersch (2005)
FEMWATER	Lin *et al.* (1997)
HEATFLOW	Frind (1982a)
HST3D	Kipp (1987)
HYDROGEOSPHERE	Therrien *et al.* (2004)
METROPOL	Leijnse and Hassanizadeh (1989)
MITSU3D	Ibaraki (1998)
MOCDENSE	Sanford and Konikow (1985)
MODHMS	HydroGeoLogic Inc. (2003)
NAMMU	Herbert *et al.* (1988)
PHT3D	Post and Prommer (2007)
SEAWAT-2000	Langevin and Guo (2006)
SUTRA	Voss (1984); Voss and Provost (2003)
SWIFT	Reeves *et al.* (1986)
TOUGH2	Oldenburg and Pruess (1995)
VapourT	Mendoza and Frind (1990)
VARDEN	Kuiper (1983, 1985); Kontis and Mandle (1988)

7.5 MODELLING VARIABLE DENSITY FLOW PHENOMENA

An excellent treatment on this subject was presented in a review article by Diersch and Kolditz (2002) and the reader is referred to that material for an exhaustive discussion on this topic. The limited number of analytical solutions for variable density flow problems creates a heavy dependence on numerical simulators. A large and growing number of numerical simulators now exist for the simulation of variable density flow phenomena, as illustrated in Table 7.2. Several of these codes will be demonstrated in later sections of this chapter where it will be seen that there are numerous formulations that are used for governing equations and their numerical solution. In this section, we demonstrate the approach to variable density flow modelling by using the SUTRA (Saturated–Unsaturated TRAnsport) numerical model (Voss, 1984 and Voss and Provost, 2003) to demonstrate key issues in variable density flow simulation. SUTRA is one of the earliest codes that emerged for the simulation of variable density flow phenomena, and is in wide use today. A summary of the history of SUTRA and its use may be found in Voss (1999).

7.5.1 Physical processes, governing equations and numerical solutions

Using the example of the SUTRA code (Voss, 1984), we demonstrate the general form of the governing equations used by variable density numerical models. The SUTRA code is a numerical solver of two general balance equations for variable density single-phase saturated–unsaturated flow and single-species (solute or energy) transport based on Bear (1979). Further details on this code may be found in Voss (1984). The summary presented below is originally found in the texts by Voss (1999) and Voss and Provost (2003).

SUTRA PROCESSES
Voss and Provost (2003) provide a detailed discussion on this subject. Simulation using SUTRA is in two or three spatial dimensions. A pseudo-3D quality is provided for 2D, in that the thickness of the 2D region in the third direction may vary from point to point. A 2D simulation may be done either in the areal plane or in a cross-sectional view. The 2D spatial coordinate system may be either Cartesian (x, y) or radial–cylindrical (r, z). Areal simulation is usually physically unrealistic for variable density fluid and for unsaturated flow problems. The 3D spatial coordinate system is Cartesian (x, y, z).

Groundwater flow is simulated through numerical solution of a fluid mass-balance equation. The groundwater system may be either saturated, or partly or completely unsaturated. Fluid density may be constant, or vary as a function of solute concentration or fluid temperature.

SUTRA tracks the transport of either solute mass or energy in flowing groundwater through a unified equation, which represents the transport of either solute or energy. Solute transport is simulated through numerical solution of a solute mass-balance equation where solute concentration may affect fluid density. The single solute species may be transported conservatively, or it may undergo equilibrium sorption (through linear, Freundlich, or Langmuir isotherms). In addition, the solute may be produced or decay through first- or zero-order processes. Energy transport is simulated through numerical solution of an energy-balance equation. The solid grains of the aquifer matrix and fluid are locally assumed to have equal temperature, and fluid density and viscosity may be affected by the temperature.

Most aquifer material, flow and transport parameters may vary in value throughout the simulated region. Sources and boundary conditions of fluid, solute and energy may be specified to vary with time or may be constant. SUTRA dispersion processes include diffusion and two types of fluid velocity-dependent dispersion. The standard dispersion model for isotropic media assumes direction-independent values of longitudinal and transverse dispersivity. A flow-direction-dependent dispersion process for anisotropic media is also provided. This process assumes that longitudinal and transverse dispersivities vary depending on the orientation of the flow direction relative to the principal axes of aquifer permeability.

SUTRA GOVERNING EQUATIONS

The general fluid mass balance equation that is usually referred to as the 'groundwater flow model' is

$$\left(S_w\rho S_{0p} + \varepsilon\rho\frac{\partial S_w}{\partial p}\right)\frac{\partial p}{\partial t} + \left(\varepsilon S_w\frac{\partial\rho}{\partial U}\right)\frac{\partial U}{\partial t} - \nabla\cdot\left[\left(\frac{\mathbf{k}k_r\rho}{\mu}\right)\cdot(\nabla p - \rho\mathbf{g})\right] = Q_p \quad (7.5)$$

where:

S_w is the fractional water saturation [1],

ε is the fractional porosity [1],

p is the fluid pressure [kg/m·s^2],

t is time [s],

U is either solute mass fraction, C [kg$_{solute}$/kg$_{fluid}$], or temperature, T [°C],

$\mathbf{k}$ is the permeability tensor [m^2],

k_r is the relative permeability for unsaturated flow [1],

μ is the fluid viscosity $\mu(T)$ [kg/ m·s],

$\mathbf{g}$ is the gravity vector [m/s^2],

Q_p is a fluid mass source [kg/m^3·s].

Additionally,

ρ is the fluid density [kg/m^3], expressed as

$$\rho = \rho_0 + \frac{\partial\rho}{\partial U}(U - U_0) \quad (7.6)$$

where:

U_0 is the reference solute concentration or temperature,

ρ_0 is fluid density at U_0,

$\frac{\partial\rho}{\partial U}$ is the density change with respect to U (assumed constant),

and

S_{0p} is the specific pressure storativity [kg/m·s^2]$^{-1}$

$$S_{0p} = (1 - \varepsilon)\alpha + \varepsilon\beta \quad (7.7)$$

where:

α is the compressibility of the porous matrix [kg/m·s^2]$^{-1}$,

β is the compressibility of the fluid [kg/m·s^2]$^{-1}$.

Fluid velocity, $\mathbf{v}$ [m/s], is given by

$$\mathbf{v} = -\left(\frac{\mathbf{k}k_r\rho}{\varepsilon S_w\mu}\right)\cdot(\nabla p - \rho\mathbf{g}) \quad (7.8)$$

The solute mass balance and the energy balance are combined in a unified solute/energy balance equation usually referred to as the 'transport model':

$$[\varepsilon S_w\rho c_w + (1 - \varepsilon)\rho_s c_s]\frac{\partial U}{\partial t} + \varepsilon S_w\rho c_w\mathbf{v}\cdot\nabla U -$$
$$\nabla\cdot\{\rho c_w[\varepsilon S_w(\sigma_w\mathbf{I} + \mathbf{D}) + (1 - \varepsilon)\sigma_s\mathbf{I}]\cdot\nabla U\} \quad (7.9)$$
$$= Q_p c_w(U^* - U) + \varepsilon S_w\rho\gamma_1^w U + (1 - \varepsilon)\rho_s\gamma_1^s U_s + \varepsilon S_w\rho\gamma_0^w + (1 - \varepsilon)\rho_s\gamma_0^s$$

where:

c_w is the specific heat capacity of the fluid [J/kg·°C],

c_s is the specific heat capacity of the solid grains in porous matrix [J/kg·°C],

ρ_s is the density of the solid grains in porous matrix [kg/m^3],

σ_w is the diffusivity of energy or solute mass in the fluid (defined below),

σ_s is the diffusivity of energy or solute mass in the solid grains (defined below),

$\mathbf{I}$ is the identity tensor [1],

$\mathbf{D}$ is the dispersion tensor [m^2/s],

U^* is the temperature or concentration of a fluid source (defined below),

γ_1^w is the first-order solute production rate [s^{-1}],

γ_1^s is the first-order sorbate production rate [s^{-1}],

γ_0^w is an energy source [J/kg·s] or solute source [kg$_{solute}$/kg$_{fluid}$·s] within the fluid (zero-order production rate),

γ_0^s is an energy source [J/kg·s] or solute source [kg$_{solute}$/kg$_{fluid}$·s] within the solid (zero-order production rate).

In order to cause the equation to represent either energy or solute transport, the following definitions are made:

for energy transport

$$U \equiv T, \quad U^* \equiv T^*, \quad \sigma_w \equiv \frac{\lambda_w}{\rho c_w}, \quad \sigma_s \equiv \frac{\lambda_s}{\rho c_w}, \quad \gamma_1^w \equiv \gamma_1^s \equiv 0;$$

for solute transport

$$U \equiv C, \quad U_s \equiv C_s, \quad U^* \equiv C^*, \quad \sigma_w \equiv D_m,$$
$$\sigma_s \equiv 0, \quad c_s \equiv \kappa_1, \quad c_w \equiv 1;$$

where:

λ_w is the fluid thermal conductivity [J/s·m·°C],

λ_s is the solid-matrix thermal conductivity [J/s·m·°C],

T is the temperature of the fluid and solid matrix [°C],

C is the concentration of solute (mass fraction) [kg$_{solute}$/kg$_{fluid}$],

T^* is the temperature of a fluid source [°C],

C^* is the concentration of a fluid source [kg$_{solute}$/kg$_{fluid}$],

C_s is the specific concentration of sorbate on solid grains [kg$_{solute}$/kg$_{grains}$],

D_m is the coefficient of molecular diffusion in porous medium fluid [m^2/s],

κ_1 is a sorption coefficient defined in terms of the selected equilibrium sorption isotherm (see Voss, 1984, for details).

The dispersion tensor is defined in the classical manner:

$$D_{ii} = |\mathbf{v}|^{-2}\left(d_L v_i^2 + d_T v_j^2\right) \qquad (7.10a)$$

$$D_{ij} = |\mathbf{v}|^{-2}(d_L - d_T)\left(v_i v_j\right) \qquad (7.10b)$$

for $i = x, y$ and $j = x, y$ but $i \neq j$

with:

$$d_L = \alpha_L|\mathbf{v}| \qquad (7.10c)$$

$$d_T = \alpha_T|\mathbf{v}| \qquad (7.10d)$$

where:

d_L is the longitudinal dispersion coefficient [m²/s],
d_T is the transverse dispersion coefficient [m²/s],
α_L is the longitudinal dispersivity [m],
α_T is the transverse dispersivity [m].

A useful generalisation of the dispersion process is included in SUTRA allowing the longitudinal and transverse dispersivities to vary in a time-dependent manner at any point *depending on the direction of flow*. The generalisation for two-dimensional flow and transport is:

$$\alpha_L = \frac{\alpha_{L\max}\alpha_{L\min}}{\alpha_{L\min}\cos^2\theta_{kv} + \alpha_{L\max}\sin^2\theta_{kv}} \qquad (7.11a)$$

$$\alpha_T = \frac{\alpha_{T\max}\alpha_{T\min}}{\alpha_{T\min}\cos^2\theta_{kv} + \alpha_{T\max}\sin^2\theta_{kv}} \qquad (7.11b)$$

where:

$\alpha_{L\max}$ is the longitudinal dispersivity for flow in the *maximum permeability direction* [m],
$\alpha_{L\min}$ is the longitudinal dispersivity for flow in the *minimum permeability direction* [m],
$\alpha_{T\max}$ is the transverse dispersivity for flow in the *maximum permeability direction* [m],
$\alpha_{T\min}$ is the transverse dispersivity for flow in the *minimum permeability direction* [m],
θ_{kv} is the angle from the *maximum permeability direction to the local flow direction* [degrees].

(See Voss and Provost, 2003, for three-dimensional direction-dependent dispersivity relations.) Relations (7.11a,b) provide dispersivity values that are equal to the square of the radius of an ellipse that has its major axis aligned with the maximum permeability direction. The radius direction is the direction of flow.

In the case of variable density saturated flow with non-reactive solute transport of total dissolved solids or chloride, and with no internal production of solute, equations (7.5), (7.8) and (7.9) are greatly simplified. The fluid mass balance is:

$$\left(\rho S_{0p}\right)\frac{\partial p}{\partial t} + \left(\varepsilon\frac{\partial\rho}{\partial U}\right)\frac{\partial U}{\partial t} - \nabla\cdot\left[\left(\frac{\mathbf{k}\rho}{\mu}\right)\cdot(\nabla p - \rho\mathbf{g})\right] = Q_p \qquad (7.12)$$

The fluid velocity is given by:

$$\mathbf{v} = -\left(\frac{\mathbf{k}\rho}{\varepsilon\mu}\right)\cdot(\nabla p - \rho\mathbf{g}) \qquad (7.13)$$

and the solute mass balance is:

$$(\varepsilon\rho)\frac{\partial U}{\partial t} + \varepsilon\rho\mathbf{v}\cdot\nabla C - \nabla\cdot[\varepsilon\rho(D_m\mathbf{I} + \mathbf{D})\cdot\nabla U] = Q_p(C^* - C) \qquad (7.14)$$

SUTRA NUMERICAL METHODS

Again, using the example code SUTRA, we demonstrate how one popular variable density numerical model solves the above governing equations. A previous summary of the discussion by Voss (1999) on this subject is repeated below.

The numerical technique employed by the SUTRA code to solve the above equations uses a modified two-dimensional Galerkin finite element method with bilinear quadrilateral elements. Solution of the equations in the time domain is accomplished by the implicit finite difference method. Modifications to the standard Galerkin method that are implemented in SUTRA are as follows. All non-flux terms of the equations (e.g. time derivatives and sources) are assumed to be constant in the region surrounding each node (cell) in a manner similar to integrated finite differences. Parameters associated with the non-flux terms are thus specified nodewise, while parameters associated with flux terms are specified element-wise. This achieves some efficiency in numerical calculations while preserving the accuracy, flexibility and robustness of the Galerkin finite element technique. Voss (1984) gives a complete description of these numerical methods as used in the SUTRA code.

An important modification to the standard finite element method that is required for variable density flow simulation is implemented in SUTRA. This modification provides a velocity calculation within each finite element based on *consistent* spatial variability of pressure gradient, ∇p, and buoyancy term, $\rho\mathbf{g}$, in Darcy's law, equation (7.13). Without this 'consistent velocity' calculation, the standard method generates spurious vertical velocities everywhere there is a vertical gradient of concentration within a finite element mesh, even with a hydrostatic pressure distribution (Voss, 1984; Voss and Souza, 1987). This has critical consequences for simulation of variable density flow phenomena. For example, the spurious velocities make it impossible to simulate a narrow transition zone between freshwater and seawater with the standard method, irrespective of how small a dispersivity is specified for the system.

The two governing equations, fluid mass balance and solute mass (or energy) balance, are solved sequentially on each iteration or time step. Iteration is carried out by the Picard method with linear half-time-step projection of non-linear coefficients on the first iteration of each time step. Iteration to the solution for each time step is optional. Velocities required for solution of the transport equation are the result of the flow equation solution (i.e. pressures) from the previous iteration (or time step for non-iterative solution).

SUTRA is coded in a modular style, making it convenient for sophisticated users to modify the code (e.g. replace the existing direct banded-matrix solver) or to add new processes. Addition of new terms (i.e. new processes) to the governing equations is a straightforward process usually requiring changes to the code in very few lines. Also to provide maximum flexibility in applying the code, all boundary conditions and sources may be time-dependent in any manner specified by the user. To create time-dependent conditions, the user must modify subroutine BCTIME. In addition, any desired unsaturated functions may be specified by user-modification of subroutine UNSAT.

One numerical feature of the code is not often recommended for practical use: upstream weighting. Upstream weighting in transport simulation decreases the spatial instability of the solution, but only by indirectly adding additional dispersion to the system. For full upstream weighting, the additional longitudinal dispersivity is equivalent to one-half the element length along the flow direction in each element.

7.5.2   Testing variable density aspects of variable density numerical codes

Diersch and Kolditz (2002) provide a comprehensive summary of model test cases and benchmarks, including details on the successes and limitations of each case. As summarised from that article, the major test cases that are used for numerical simulators currently include the following.

1. *The hydrostatic test*: This type of benchmark is quite simple, but is very instructive. It is a test of the velocity consistency under hydrostatic and sharp density transition conditions as originally proposed by Voss and Souza (1987): a closed rectangular region containing freshwater above highly saline water, separated by a horizontal transition zone one element wide and with a hydrostatic pressure distribution (Voss and Souza, 1987). No simulated flow may occur in the cross-section under transient or steady-state conditions.

2. *The Henry seawater-intrusion problem*: This test case has been performed on coarse meshes (Voss and Souza, 1987) and on a very fine mesh (Segol, 1993). Note that until Segol (1993) improved Henry's approximate semi-analytical solution (Henry, 1964), this test was merely an inter-comparison of numerical code results. No published results had matched the Henry solution. Segol's result provides a true analytic check on a variable density code. Some confusion has existed in the literature over the correct choice of diffusion

to use in the model test case and apparently two diffusivity values have been employed that differ by a factor of porosity. In addition to this problem, the Henry problem has some other deficiencies. An unrealistically large amount of diffusion is often introduced which results in a wide transition zone. This assists with the numerical solution, making the solution smooth and less problematic. However, as pointed out by Diersch and Kolditz (2002), the Henry problem is not appropriate for testing purely density-driven flow systems. This is because a large component of the flow system response is actually driven by forced convection and not free convection. As a result, additional benchmark tests are necessary to test numerical models for free convection problems (such as the Elder problem) and for cases with very narrow transition zones (such as the salt-dome problem).

3. *The Elder natural convection problem* (Voss and Souza, 1987): This is an unstable transient problem consisting of dense fluid circulating downwards under buoyancy forces from a region of high specified concentration along the top boundary. The Elder problem serves as an example of free convection phenomena, where the bulk fluid flow is driven purely by fluid density differences. Elder (1967a, 1967b) presented both experimental results in a Hele-Shaw cell and numerical simulations concerning the thermal convection produced by heating a part of the base of a porous layer (what we often now call the 'short heater problem'). The Elder problem is a very popular benchmark case and its convective flow circulation patterns (see Figure 7.3) are still the subject of scientific discussion. No exact or qualified measurements exist for the Elder problem and as such the test case relies on an inter-code comparison between numerical simulators. The main issue that has been discussed widely in relation to this problem is the nature of both the number of fingers and the central upwelling or downwelling patterns, which are now recognised to be highly dependent on the nature of the grid discretisation (i.e. coarse, fine, extremely fine grids all lead to very different spatio-temporal patterns of behaviour). It is likely that convergence has been achieved at extremely fine grid resolutions (15 000–1 000 000 nodes) as reported in Diersch and Kolditz (2002).

4. *The HYDROCOIN salt-dome problem* (e.g. see Konikow *et al.*, 1997): This is a steady-state and transient problem, initially consisting of a simple forced freshwater flow field sweeping across a salt dome (specified concentration located along the central third of the bottom boundary). This generates a separated region of brine circulating along the bottom. The nature of the circulation patterns has been shown to be sensitive to the way in which the bottom boundary condition is represented and implemented, and

(a)   (b)

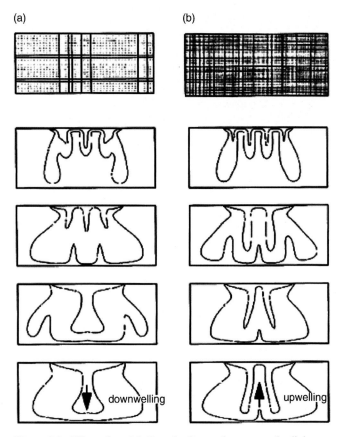

Figure 7.3. Effect of spatial discretisation on the computed salinity distribution at 4, 10, 15 and 20 years simulation time. Positions of the 20% and 60% isochlors are shown for: (a) coarse mesh (1170 nodes, 1100 quadrilateral elements), comparable with the discretisation used originally by Elder (1967a, 1967b) and Voss and Souza (1987), (b) fine mesh (10 108 nodes, 9900 quadrilateral elements). It is clear that the nature of the central upwelling and downwelling is very sensitive to grid discretisation. Source: Diersch and Kolditz (2002).

the importance choice of mechanical dispersion. These are seen to affect the circulation patterns that arise in the system.

5. *The salt lake problem*: First introduced by Simmons *et al.* (1999), this is a very complicated, transient system in which an evaporation- and density-driven fingering process evolves downwards in an area of upward-discharging groundwater in a Hele-Shaw cell experiment (see Figure 7.4). Numerical results (see Figure 7.5) are compared with laboratory measurements. There has been much discussion on this test case in recent literature. In particular, the number of fingers and their temporal evolution appear to be intimately related to the choice of numerical discretisation and the numerical solution procedure. Interestingly, and paradoxically, several authors have now found that better agreement between numerical results and experimental results is obtained using a coarser, rather than finer, mesh (Mazzia *et al.*, 2001; Diersch and Kolditz, 2002). Numerical grid

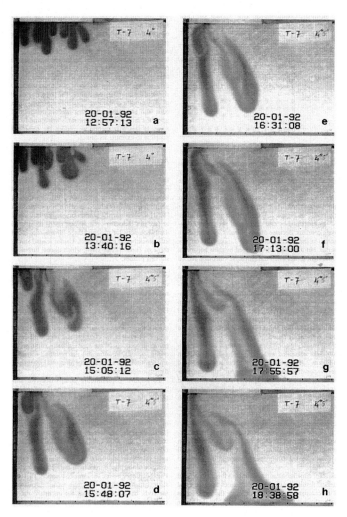

Figure 7.4. Developmental stages of unstable plume behaviour in Hele-Shaw cell laboratory experiment. The effective Rayleigh number for this system is $Ra = 4870$. Elapsed times (a, b, . . ., h) given in dimensionless times are $T = 2.02, 4.03, 8.00, 10.01, 12.02, 13.98, 15.99, 18.00$, measured from a virtual starting time of 12 hours and 15 minutes (the actual experimental starting time was 11 hours and 32 minutes). One unit in dimensionless time is equivalent to 21.4 minutes real time (after Wooding *et al.*, 1997b; Simmons *et al.*, 1999).

convergence has not been achieved for this problem. The higher Rayleigh number of this test case problem (about 10 times larger than for the Elder problem) is clearly in a range where oscillatory convection flow regimes are expected. In comparison to the Elder problem, the salt lake problem is much more complicated. Grid convergence for the Elder problem required extremely fine numerical resolution, and we therefore expect this to be increased further again for the case of the salt lake problem. The major challenge here is that this problem is extremely dynamic and that it is unstable. Moreover, numerical perturbations and dispersion can both trigger fingering and control associated dispersion (creating either fatter or skinnier fingers!) and these are all intimately related

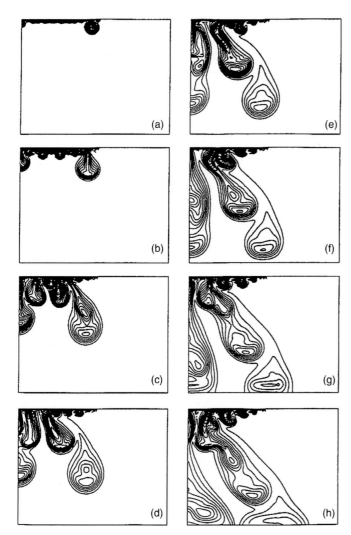

Figure 7.5. Developmental stages of unstable plumes in the SUTRA solute transport model. Contours represent the concentration of the plumes. Contour interval is 2000 mg l^{-1}. The effective Rayleigh number for this system is $Ra = 4870$. Elapsed times (a, b, . . ., h) given in dimensionless times are $T = 2.01, 4.02, 8.03, 10.05, 12.06, 14.06, 16.07, 18.08$, measured from start of numerical simulation. One unit in dimensionless times is equivalent to 21.4 minutes real time. Source: Simmons *et al.* (1999).

to both the patterns of finger formation as well as their subsequent growth/decay processes. It is clear that the numerical simulation of such phenomena is extremely challenging and appears to be currently unresolved.

6. *The saltpool problem*: The saltpool problem was introduced by Johannsen *et al.* (2002). It represents a three-dimensional saltwater upconing process in a cubic box under the influence of both density-driven flow and hydrodynamic dispersion. A stable layering of saltwater below freshwater is considered in time for both a low-density case (1% salt mass fraction) and a high-density case (10% salt mass fraction). A laboratory experiment was used as part of the

model testing process. As freshwater sweeps over the stable saltier water at the bottom of the cube, the breakthrough curves at the outlet point are measured. The position of the saltwater–freshwater interface was measured using nuclear magnetic resonance (NMR) techniques. Numerous authors have attempted to model this problem and have achieved variable success. The challenge appears to lie in the small dispersivity values and the large density contrasts, particularly in the high-density cases. The experiment and test case are resolved for low-density contrasts and for early time behaviour. Some questions do remain about the simulation of high-density contrasts and late time behaviour. Previous studies have emphasised the need for a consistent velocity approximation in the accurate simulation of the saltpool problem. As noted by Diersch and Kolditz (2002), the saltpool experiment does provide reliable quantitative results for a three-dimensional saltwater mixing process under density effects which were not available prior to its introduction. It is seen that extreme mesh refinement is required in order to accurately simulate the problem.

More recently than the publication of the work by Diersch and Kolditz (2002), a very new test case suite has been proposed by Weatherill *et al.* (2004) for two-dimensional convection and by Voss *et al.* (2010) for three-dimensional convection that have an exact and irrefutable solution based upon an analytic stability theory. This is:

7. *Convection in infinite, finite and inclined porous layers*: A body of work was developed in the 1940s, originally by Horton and Rogers (1945) and independently by Lapwood (1948), to examine the convective stability of an infinite layer with an unstable density configuration applied across it. The work has an exact analytical solution for both the onset of convection (a critical Rayleigh number of $4\pi^2$) and the wavelength of the convection pattern. Unlike all previous test cases, the availability of analytically based stability criteria offers an excellent way of testing a numerical code which does not rely on a comparison with other numerical simulators. Additionally, these stability criteria have been applied in traditional convection problems in the field of fluid mechanics for many decades both within numerical and laboratory-based frameworks. From a scientific viewpoint they are well understood, robust and irrefutable. They are therefore ideal for testing numerical models. In their study, Weatherill *et al.* (2004) compared numerical results using SUTRA with these previous studies and showed that SUTRA results (for both the onset conditions for instability and its resultant wavelength) were in excellent agreement with the traditional stability criteria. Numerical solutions to extensions of this original problem, namely cases of finite layers and inclined layers (where analytical stability criteria also exist), were seen to be in excellent agreement

with stability criteria and previous results reported in the traditional fluid-mechanics literature. Given the availability of exact and irrefutable stability criteria for these 'box problems', it is expected that they will become commonplace in future numerical model testing. They also have solutions that were extended into 3D benchmark test cases (Voss, *et al.* 2010).

7.5.3 Future challenges for numerical modelling

Simmons (2005) and Diersch and Kolditz (2002) highlighted some current challenges that face the field of variable density groundwater flow and these clearly have critical implications for the modelling of such phenomena. Simmons (2005) identified a number of emerging challenges and suggested that there is little, if any, evidence in current literature that points to the future extinction of this field of hydrogeology. Future drivers for both research and practical application in this area will drive numerical modelling needs.

Some future drivers for variable density flow research, and hence modelling, identified by Simmons (2005) included, among others:

(i) the need to better understand the relationship between variable density flow phenomena and dispersion;

(ii) better geological constraints on variable density flow analyses using lineaments, sedimentary facies data and structural properties of fractured rock aquifers;

(iii) improving the resolution of geophysical, geoprospecting and remote sensing tools for non-invasive characterisation of dense plume phenomena and heterogeneity;

(iv) linking the fields of tracer and isotope hydrogeology and variable density flow phenomena;

(v) double-diffusive and multiple species transport problems in variable density flow phenomena;

(vi) links between climate change phenomena and the response of variable density flow phenomena, such as sea level and coastline position change, and the impact of transgression and/or regression cycles on aquifer salinisation will be critical;

(vii) the gas–liquid phase chemistry of carbon sequestration processes and its efficiency, safety and long-term viability as a storage option will necessarily involve an understanding of variable density transport dynamics with multiple phases;

(viii) links between variable density flow phenomena and a number of hydro-ecological applications (e.g. how does variable density flow affect baseflow accessions and other surface–groundwater interactions, or salt budge profiles under vegetation, or the spatio-temporal distribution of stygofauna and biota in subsurface groundwater ecosystems?).

Some of the above issues point to the need for more detailed and accurate coupling of surface water, vadose zone and groundwater models – and which will drive an inherent increase in the complexity of the modelling approach. The numerous future possibilities outlined above suggest both short-term and long-term prognoses for variable density flow research and its applications are very good. It is also logical to conclude that there are many issues that remain largely unexplored and poorly understood in the numerical modelling of variable density phenomena.

Some major future challenges for the field of variable density flow phenomena are summarised below. Here, those matters of particular importance to the modelling of variable density flow systems are highlighted. This extended analysis is based on the treatment presented by Simmons (2005). Some major challenges identified in that study include:

1. *Large mix of spatial and temporal scales*: Variable density flow phenomena may be triggered, grow and decay over a very large mix of different spatial and temporal scales. The particular challenge lies in the fact that information on very small spatial scales and short time-scales is needed to feed into long time and large spatial scale processes and simulations. Measuring field scale parameters across this large range of spatial and temporal scales as input for modelling approaches is a major challenge. Additionally, many current groundwater systems are highly transient and steady assumptions may be an oversimplification, especially where solute transport is concerned. For example, tidal oscillations and sea-level rise create the need for transient analyses in seawater-intrusion phenomena. In unstable problems, the onset and growth/decay of instabilities requires a truly transient analysis of the system. Furthermore, the source of dense plumes (e.g. leachate from a landfill) is often represented by simplified constant head, flux or concentration boundary conditions and yet the style of loading is expected to be important (Zhang and Schwartz, 1995). There are challenges in better understanding how transient boundary conditions will affect variable density flow phenomena (e.g. continuous or intermittent sources).

2. *Heterogeneity and dispersion*: It is only more recently that heterogeneity has been considered in the study of dense plume migration (e.g. Schincariol and Schwartz, 1990; Schincariol *et al.*, 1997). Schincariol *et al.* (1997) explain that heterogeneity in hydraulic properties can perturb flow over many length scales (from slight differences in pore geometry to larger heterogeneities at the regional scale), triggering instabilities in density-stratified systems. This is particularly true for physically unstable situations but is also important in other situations. For example, the mixing process along a saltwater–freshwater interface can be significantly modified by non-uniform velocities. This can lead to local physical instabilities resulting in a corrugated interface that at larger scales is interpreted as increased dispersion. Modelling narrow transition zones and low dispersion

situations is challenging, most often requiring very fine numerical grids and increased computational power. Current research is suggesting that heterogeneity may serve as a physical perturbation in fingering processes and is therefore a critical feature controlling the onset, growth and/or decay in variable density flow processes. For example, low-permeability lenses can effectively dampen instability growth or even completely stabilise a weakly unstable plume. Dense plume migration, particularly at high-density differences and in highly heterogeneous distributions, is therefore not easily amenable to prediction (Schincariol et al., 1997; Simmons et al., 2001; Nield and Simmons, 2006). While some studies have examined dense plume migration in fractured rock using numerical models (Shikaze et al., 1998; Graf and Therrien, 2005), there is still a need to explore how more complex and realistic fracture geometries affect variable density flow processes and to better understand the links with macroscopic dispersion (e.g. Welty et al., 2003; Kretz et al., 2003; Schotting et al., 1999).

3. *Fluids, solutes and fluid–matrix interaction*: Previous studies have typically been limited to single species conservative transport but some have examined heat and solute transport simultaneously within thermohaline convection regimes (see Diersch and Kolditz, 2002, for a discussion on this topic in their exhaustive review of variable density modelling) which may be important in understanding many processes including mineralisation and ore formation in sedimentary basins, geothermal extraction processes and flow near salt domes. The simultaneous interaction of heat and solute creates challenges of its own and this complexity is exemplified where multiple species with differing diffusivities interact. In a subset of these 'double-diffusive' processes, instabilities can occur even where the net density is increasing downwards. Here, it is the difference in the diffusivities of the heat and solute that drives transport. It should be realised that the simultaneous interaction of heat and solute creates many challenges of its own and that this complexity will be exemplified where multiple species with differing diffusivities interact. Multiple species studies (beyond the two species heat and salt) are fairly limited in groundwater literature. However, more recent studies (e.g. Zhang and Schwartz, 1995) suggest that the chemical composition and reactive character of a plume can greatly influence plume dynamics. Other studies have begun to examine how chemical reactions can be coupled with density-dependent mass transport. For example, Freedman and Ibaraki (2002) incorporated equilibrium reactions for the aqueous species, kinetic reactions between the solid and liquid phases, and full coupling of porosity and permeability changes that result from precipitation and dissolution reactions in porous media and showed that complex concentration distributions result in the variable density flow system. Recently, Ophori (1998) suggested that the variable viscosity nature of fluids is typically ignored in variable density groundwater flow problems and suggest that plume migration pathways and rates are inaccurately predicted in the absence of the variable-viscosity relationship. Other studies have suggested that dense plume migration is significantly affected by the extent of porous medium saturation. Thorenz et al. (2002) and Boufadel et al. (1997) demonstrate that significant lateral flows and coupled density-driven flow may take place in the partially saturated region above the water table and at the interface between the saturated and partially saturated zones. All these studies suggest that further investigation in relation to multiple species transport, non-conservative solute transport, fluid–matrix interactions and a more complex representation of the chemical and physical properties of the dense fluid than has previously been assumed is warranted. The coupling of variable density flow phenomena with more complex chemical and fluid property characterisation is an area that is still in its infancy and warrants further exploration. As additional complexity is added to the numerical simulator in this area, there will be a need to identify new test cases that actually test those newly added features of the code.

4. *Numerical modelling*: In general, variable density flow problems do not lend themselves easily to analytical study except under the most simplified of conditions. However, a number of numerical codes such as those outlined in Table 7.2 now exist to simulate variable density flow phenomena. There has been good success in simulating variable density flow at low to moderate density contrasts and in benchmarking the codes that are used. As was seen in the analysis of test case problems, numerical modelling efforts were most successful in the cases of stable problems (e.g. the Henry saltwater intrusion case), but increasing problems are encountered in the case of unstable problems (e.g. confusion over downwelling/upwelling patterns and the massive computational power required for grid convergence in the case of the Elder problem, and the currently unresolved nature of the salt lake problem where grid convergence has not currently been achieved). But it appears that even seemingly simple matters in numerical modelling remain unresolved and appear to be currently overlooked.

It is clear that variable density problems increase in modelling complexity as one moves from stable to unstable configurations. The modelling complexity and computational demand is compounded for large density differences in both stable (e.g. the saltpool problem) and unstable problems (e.g. the salt lake problem) and in cases where physical dispersion is very small.

Additionally, artefacts of the numerical scheme itself can mimic physically realistic but perhaps completely inaccurate results (e.g. dispersion enhancing the transition zones in stable problems, or enhancing the width and reducing their speed of unstable fingers). The problem is compounded at much higher (unstable) density differences and hence Rayleigh numbers where the physical problem itself is characterised by chaotic and oscillatory regimes such as that seen in the proposed 'salt lake problem' (Simmons *et al.*, 1999). What is particularly problematic in the simulation of such unstable phenomena is that small differences in dispersion parameters or spatial and temporal discretisation can cause very different types of instabilities to be generated (Diersch and Kolditz, 2002). The nature of physical perturbations (cf. numerical perturbations) is almost impossible to measure and simulate. When using standard numerical approaches, one may attempt to control numerical perturbations by reducing numerical errors through the use of very fine meshes and small time-steps but the nature of what physical perturbation is present in the system in order to generate particular patterns of instabilities is never known. It is almost impossible to resolve, therefore, what are numerical artifacts and what are physical realities.

Furthermore, as noted by Diersch and Kolditz (2002), one must then ask what perturbation is no longer significant for a convection process, and what error in the numerical scheme can be tolerated. These authors note that there are two possible dangers in simulating such highly non-linear convection problems within a numerical framework. First, a scheme which is only conditionally stable, but of higher accuracy, is prone to create non-physical perturbations due to small perturbations in the solutions unless the element size and the time-step are sufficiently small. Conversely, an unconditionally stable numerical scheme but with lower accuracy, smooths out physically induced perturbations, unless the element size and the time-step are sufficiently small. It is clear that a very fine spatial and temporal resolution are essential if high-density contrast convection phenomena are to be modelled. One may conclude that the simulation of unstable phenomena, particularly highly unstable cases, remains a challenge for further research.

In addition to the lack of control over numerical perturbations and the inability to quantify the physical perturbation in real convective systems, it is also questionable whether oscillatory or chaotic regimes such as those encountered at high Rayleigh numbers can be simulated and predicted by a deterministic modelling approach. In more general terms, it can be seen that variable density flow problems are currently limited by the need to simulate large density contrast and low dispersion systems. Both are massively computationally demanding. As noted by Diersch and Kolditz (2002), large-scale variable density simulations have an inherent need for expensive numerical meshes. Some of the most promising developments in the numerical

modelling of real field-based problems include adaptive techniques, unstructured meshes and parallel processing. Diersch and Kolditz (2002) note that adaptive and unstructured irregular geometries are prone to uncontrollable perturbations and that robust, efficient and accurate strategies are required, particularly for three-dimensional applications.

Given these numerical complexities and demands, a critical matter that warrants attention in modelling of such systems is the level of simplification that is possible without loss of physical information and accuracy. In many cases the science in relation to this question is ambiguous, unresolved or is simply ignored. For example, when can homogeneous equivalents for heterogeneous systems be employed? When is a 2D analysis of a 3D system permissible? What are the implications of such a simplification? The complexity–simplicity matter must be considered carefully in the modelling of any variable density flow system. In many cases, it appears that the decision is made implicitly, often with little or no justification. Without a systematic evaluation of modelling complexity and the subsequent implementation of a parsimonious modelling framework, it is apparent that the implication of taking either a complex or simple approach may often not be properly understood. Finally, and perhaps most importantly, there is also a real challenge for greatly improved field measurements and observations for field truthing of models, minimising their non-uniqueness and to inspire greater confidence in model predictive output.

7.6 APPLICATIONS AND CASE STUDIES

In this section, we demonstrate a range of case studies and applications where variable density flow is considered important. These examples include saltwater intrusion and tidally induced phenomena, salinisation associated with transgression and regression cycles, variable density processes in aquifer storage and recovery, flow and transport modelling in fractured rock aquifers, and, chemically reactive transport modelling. Many of these examples are relevant to arid zone areas but importantly extend beyond them. These applications and case studies are designed to illustrate important areas where the field of variable density flow modelling is developing. The applications and case studies are designed to be illustrative rather than exhaustive.

7.6.1 Seawater intrusion and tidally induced phenomena

BACKGROUND
Seawater intrusion (or saltwater intrusion) is the encroachment of saline waters into zones previously occupied by fresh groundwater. Under natural conditions, hydraulic gradients in coastal

aquifers are towards the sea and a stable interface between the discharging groundwater and seawater exists, notwithstanding climatic events or sea-level rise (e.g. Cartwright *et al.*, 2004; Kooi *et al.*, 2000). Persistent disturbances in the hydraulic equilibrium between the fresh groundwater and denser seawater in aquifers connected to the sea produce movements in the position of the seawater–freshwater interface, which can lead to degradation of freshwater resources. The impacts of seawater intrusion are widespread, and have led to significant losses in potable water supplies and in agricultural production (e.g. Barlow, 2003; Johnson and Whitaker, 2003; FAO, 1997).

Seawater intrusion is typically a complex three-dimensional phenomenon that is influenced by the heterogeneous nature of coastal sediments, the spatial variability of coastal aquifer geometry and the distribution of abstraction bores. The effective management of coastal groundwater systems requires an understanding of the specific seawater-intrusion mechanisms leading to salinity changes, including landward movements of seawater, vertical freshwater–seawater interface rise or 'up-coning', and the transfer of seawater across aquitards of multi-aquifer systems (e.g. van Dam, 1999; Sherif, 1999; Ma *et al.*, 1997). Other processes, such as relic seawater mobilisation, salt spray, atmospheric deposition, irrigation return flows and water–rock interactions, may also contribute to coastal aquifer salinity behaviour, and need to be accounted for in water resource planning and operation studies (e.g. Werner and Gallagher, 2006; Kim *et al.*, 2003; FAO, 1997).

The investigation of seawater intrusion is a challenging undertaking, and requires careful consideration of groundwater hydraulics and abstraction regimes, the influence of water density variability, the hydrochemistry of the groundwater and the inherent dynamics of coastal systems (i.e. tidal forcing, episodic ocean events such as storm surges and tsunamis, salinity and water-level variation in estuaries and other tidal waterways, among others). The challenges of seawater-intrusion investigation have led to the development of specific analytical methods and mathematical modelling tools that enhance the interpretation of field-based observations, which are invariably too sparse to provide a complete representation of seawater-intrusion processes and trends.

ANALYTICAL SOLUTIONS
The simplest analyses of seawater intrusion adopt the assumption of a sharp seawater–freshwater interface, giving rise to analytical solutions to interface geometry. The Ghyben-Herzberg approximation (developed independently by Badon–Ghyben in 1888 and by A. Herzberg in 1901; FAO, 1997) for the depth of the interface below the water table was the first such solution and was based on hydrostatic conditions, given as:

$$z = \frac{\rho_f}{\rho_s - \rho_f} h_f \qquad (7.15)$$

where z is the depth below sea level of the interface (m), ρ_f and ρ_s are the freshwater and saltwater densities (kg m^{-3}), and h_f is the height of the water table above sea level (m). Badon–Ghyben and Herzberg, working independently, showed that seawater actually occurred at depths below sea level equivalent to approximately 40 times the height of freshwater above sea level. Glover (1964) modified the solution to account for a submerged seepage face:

$$z = \frac{\rho_f}{\rho_s - \rho_f} \frac{q'}{K} + \sqrt{\frac{\rho_f}{\rho_s - \rho_f} \frac{2q'x}{K}} \qquad (7.16)$$

where q' is the discharge per unit width of aquifer (m^2 s^{-1}), K is the hydraulic conductivity (m s^{-1}) and x is the distance from the ocean boundary (m). Further advances in analytical solutions to sharp-interface problems include the works of Kacimov and Sherif (2006), Dagan and Zeitoun (1998), Naji *et al.* (1998).

Sharp-interface approaches are only applicable to settings where interface thicknesses are small in proportion to the total aquifer depth. Alternately, the dispersive nature of salt transport needs to be considered and numerical modelling approaches are therefore ultimately required in most practical applications in order to solve the coupled groundwater flow and solute transport equations.

NUMERICAL MODELS
A major limitation of the sharp-interface approach is that the movement of brackish groundwater (i.e. diluted seawater) cannot be analysed in the transition zone. Where the diffusive behaviour of solute transport is a necessary component of seawater-intrusion analyses, models are applied that adopt numerical methods to resolve the groundwater flow and solute transport equations. The computational effort of simulating density-coupled groundwater flow and solute transport is usually an imposition that influences the spatial and temporal resolution of simulations, and model grid design usually involves trade-offs between model run-times and the resolution of simulation. This is particularly important for investigations of large areas and/or where simulations are carried out in a three-dimensional domain.

The US Geological Survey's finite element SUTRA code (Voss, 1984) is arguably the most widely applied simulator of seawater intrusion and other density-dependent groundwater flow and transport problems in the world (Voss, 1999). Several other finite element codes are capable of seawater-intrusion modelling, including the popular FEFLOW model (Diersch, 2005), FEMWATER (Lin *et al.*, 1997), etc. SUTRA has been applied to a wide range of seawater-intrusion problems, ranging from regional-scale assessments of submarine groundwater discharge (SGD) (e.g. Shibuo *et al.*, 2006) to riparian-scale studies of estuarine seawater intrusion under tidal forcing effects (e.g. Werner and Lockington, 2006). Gingerich and Voss (2005)

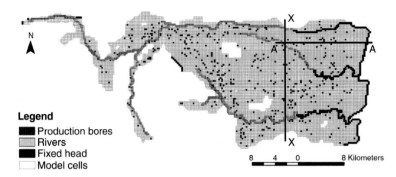

Figure 7.6. Numerical model of salt water intrusion in a regional scale 3D model, Pioneer Valley, Australia. Based upon Kuhanesan (2005) groundwater flow model. (Source: Werner and Gallagher, 2006).

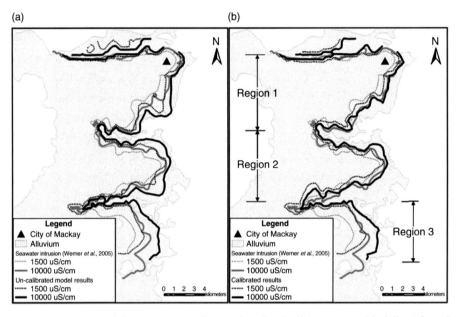

Figure 7.7. A comparison between observed salinity contours (grey lines) and predicted salinity contours (black lines) from the (a) uncalibrated model and (b) calibrated model. The coastline is not shown. (Source: Werner and Gallagher, 2006).

demonstrate the application of SUTRA to the Pearl Harbour aquifer, southern Oahu, Hawaii, in analysing the historical behaviour of the seawater front during 100 years of pumping history.

The list of seawater-intrusion simulators includes hybrids of the popular finite-difference groundwater flow code MODFLOW (Harbaugh, 2005) including SEAWAT (Langevin *et al.*, 2003) and MODHMS (HydroGeoLogic Inc., 2003). The advantage of these codes is that they can be applied to existing MODFLOW models through relatively straightforward modifications to MODFLOW input data files, thereby reducing model development effort. Further, existing MODFLOW pre- and postprocessors can be utilised. Robinson *et al.* (2007a) applied SEAWAT to the beach-scale problem of groundwater discharge under tidal forcing, and predicted the temporal behaviour of submarine groundwater discharge and the geometry of salt plumes at the coastal boundary.

Werner and Gallagher (2006) described the development of a three-dimensional seawater-intrusion model from a two-dimensional groundwater flow (MODFLOW) model, and describe some of the practical modelling problems that were encountered along the way. Practical problems included:

1. converting from 2D planar to 3D heterogeneous model frameworks while maintaining the same depth-averaged groundwater hydraulics between both 2D and 3D representations;
2. the adoption of 3D boundary conditions (i.e. the representation of partially penetrating estuaries, amongst others) and their influence on the results compared to 2D groundwater flow;
3. converting from the Dupuit assumption to variably saturated flow (and the conversion from MODFLOW to MODHMS modelling frameworks).

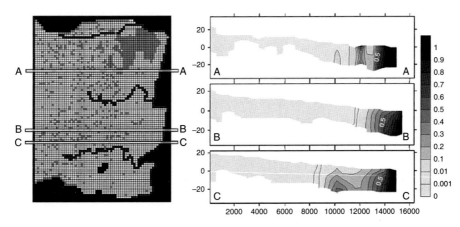

Figure 7.8. Profiles of salinity distributions (relative salinity units used, whereby 1.0 is sea water). The vertical axis scale is m AHD. (Source: Werner and Gallagher, 2006).

The modelling was used to provide an analysis of the 'susceptibility to seawater intrusion' throughout the aquifer. Susceptibility was assessed using various information sources – the modelling provided an indication of areas susceptible to future seawater intrusion by running long-term simulations (80+ years) that used predicted water resource operational scenarios. Various simulations of water resource management, including 'same as existing', 'worst case', 'severe cutbacks', were tested. Werner and Gallagher (2006) demonstrated the application of MODHMS to seawater intrusion at the regional scale, by adapting an existing hydro-dynamically calibrated MODFLOW model of the Pioneer Valley coastal plain aquifers. They used the model to demonstrate that seawater intrusion had not reached an equilibrium state, and that alternative salinisation processes (e.g. basement dissolution, relic seawater mobilisation, sea salt spray, irrigation salinity) produced observed salinity patterns in some parts of the study area. Figure 7.6 shows the original groundwater model design, Figure 7.7 provides results from the calibrated model process and finally concentration distributions are provided in Figure 7.8. The modelling suggested that the salinity in the Pioneer Valley is not all caused by classic seawater intrusion (i.e. seawater entering the aquifer from the ocean).

Robinson *et al.* (2006, 2007a, 2007b, 2007c) applied SEAWAT to the beach-scale problem of submarine groundwater discharge (SGD) and mixing under tidal forcing, and predicted the temporal behaviour of submarine groundwater discharge and salt plumes at the coastal boundary. Numerical simulations together with field results demonstrated that tides strongly influence near-shore groundwater behaviour. Tidally driven seawater circulation forms an upper saline plume in the intertidal region above discharging fresh groundwater. This upper saline plume represents an active and dynamic mixing and reaction zone which may have considerable implications for the transport of chemicals to coastal waters via SGD.

Conceptualisation and analysis of this mixing zone extends our understanding of the processes and functioning of subterranean estuaries.

A theoretical framework was developed for studying tidally dominated subterranean estuaries, in particular, how to quantify and analyse the near-shore flow and transport dynamics (Robinson *et al.*, 2007a). It was shown that the upper saline plume is associated with faster flow rates and significantly lower transit times than the dispersion zone of the saltwater wedge. Thus it represents a potentially more dynamic zone for mixing and reaction. Numerical tracer simulations illustrated that the tide strongly influences the subsurface transport of terrestrial contaminants to coastal waters by modifying the specific discharge pathway and the local geochemical conditions along this pathway. As occurs in the surface estuary, the flow patterns, salinity stratification and mixing conditions in the subterranean estuary are dependent on the relative magnitude of the inland (fresh groundwater discharge) to tidal forcing. Based on this finding, a systematic classification of subterranean estuaries was introduced.

FUTURE CHALLENGES

In recent times, seawater-intrusion modelling has become a more straightforward undertaking, given advancements in computer processing capabilities, improvements in modelling codes, and the development of user-friendly pre- and post-processors. Further, the advent and evolution of MODFLOW-based seawater-intrusion simulators facilitate a relatively uncomplicated transition from MODFLOW model to seawater-intrusion model. However, the interpretation of modelling results is still an extremely challenging task, especially where the aim of the investigation is to develop practical solutions to the mitigation and remediation of seawater intrusion.

Despite an abundance of benchmarking studies for variable density codes, the results of these have not translated into guidelines for interpreting and applying seawater-intrusion models. The reasons for this are probably because of differences in scale and in vertical-to-horizontal aspect ratios between benchmark problems and real field-scale seawater-intrusion problems. For example, both the Henry problem and the benchmark problem recently described by Goswami and Clement (2007) involve horizontal-to-vertical ratios of about 2, while the seawater-intrusion problem of Werner and Gallagher (2006) involved a horizontal-to-vertical ratio of about 500.

Another significant challenge of practical seawater-intrusion modelling is determining the degree that aquifer heterogeneities are represented in simulations of real-world systems. It is generally accepted that preferential flow paths play an important role in the behaviour of contaminant plumes; however, the degree of representation of the variability in aquifer properties has not been adequately explored for the seawater-intrusion problem. The problem of selecting the level of representation of aquifer heterogeneities needs to be considered in combination with the selection of the appropriate grid resolution.

Other potential issues include the inclusion of transient information in models (irrigation cycles, groundwater extraction regimes) that influence intrusion phenomena. All of these challenges point to the growing need for further 3D variable density and transient analyses, often covering very large spatial and temporal scales.

The additional inclusion of heterogeneity (whether it be in boundary conditions in space or time or in geologic properties of the aquifer) poses additional complexity. Resolving narrow transition zones further compounds the need for greater computational effort. Increased efficiency in numerical solvers and other numerical efficiencies will be required to meet these growing computational demands. Additionally, the large computational effort associated with seawater-intrusion modelling of regional-scale systems generally precludes the application of automated calibration techniques, such as the gradient methods adopted by standard groundwater model calibration codes like PEST (Doherty, 2004). Further research is needed to develop methods of calibrating seawater-intrusion models, and to analyse the model uncertainty of making predictions with these models.

7.6.2 Transgression–regression salinisation of coastal aquifers

IMPORTANCE OF TRANSGRESSION–REGRESSION CYCLES

Tidal forcing, seasonal changes in recharge and groundwater abstraction are key factors responsible for variability and change in coastal groundwater systems. These drivers act on human time-scales. However, when viewed on longer time-scales from centuries to millions of years, coastal zones are even more dynamic. Sea-level change, erosion and sedimentation have caused coastlines to shift back (i.e. seaward: regression) and forth (i.e. landward: transgression) over distances of up to several hundreds of kilometres. Groundwater systems responded to these conditions by adjustment of flow fields and redistribution of fresh and saline water. Although the geological changes in boundary conditions generally tend to occur at a slower pace than more recent human-induced changes or diurnal and seasonal forcing, their impact on current groundwater conditions is often enormous.

More noticeable, or even catastrophic, retreats of the coastline occur in arid regions when droughts, river diversion or irrigation schemes decrease the inputs of (fresh) water to inland seas or lakes. Well-documented cases include the Dead Sea, the Aral Sea, Lake Corangamite and Lake Chad (Nihoul *et al.*, 2004). Conversely, shoreline advance is also associated with decreased river discharge to coastal zones when decrease in sediment loads causes land loss due to erosion and subsidence. Decreased discharge of the Murray river in Australia and the Nile in Egypt are key examples (Frihy *et al.*, 2003). Besides enormously adverse economic and ecological consequences, the hydrogeological effects of such coastline shifts are of paramount importance as well. These include the relocation of discharge and recharge zones, changes in groundwater–surface interaction processes and water quality degradation. Variable density flow also plays a crucial role in these settings.

OBSERVATIONAL EVIDENCE AND CONCEPTUAL MODELS

It is essential to realise that the textbook conception of the fresh and saline groundwater distribution, which is classically conceived of as a freshwater lens overlying a wedge of saline groundwater, is seldom encountered in real field settings due to the dynamic nature of shorelines. The most conspicuous manifestations of transient effects are offshore occurrences of fresh groundwater and onshore occurrences of salt water.

Offshore fresh and brackish palaeo-groundwater

Over the years, evidence has been growing in support of the observation that sub-seafloor fresh and brackish groundwater are common features of continental shelves and shallow seas around the world. Figure 7.9 shows the known global occurrences of offshore fresh and brackish groundwater. Studies from New Jersey, USA (Hathaway *et al.*, 1979; Kohout *et al.*, 1988), New England, USA (Kohout *et al.*, 1977), Suriname, South America (Groen, 2002) and the Dutch sector of the North Sea (Post *et al.*, 2000) have shown that groundwater with salt contents between 1% and 50% of seawater occur up to 150 km from the present coastline and at depths up to 400 m below the seafloor. Additional, less conclusive cases have been reported for Port Harcourt, Nigeria, Jakarta, Indonesia, and for the Chinese Sea (Groen, 2002).

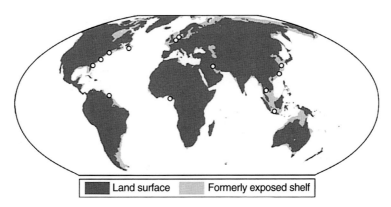

Figure 7.9. Inferred occurrences of offshore fresh to brackish palaeo-groundwater (modified from Groen, 2002) and continental shelf areas that were exposed during the last glacial maximum (25–15 ka BP).

In many instances these waters occur too far offshore to be explained by active sub-sea outflow of freshwater due to topographic drive. Moreover, lowest salinities often occur at substantial depths beneath the seafloor and are overlain by more saline pore waters, suggesting absence of discharge pathways. These waters therefore are considered palaeo-groundwaters that were emplaced during glacial periods with low sea level. During subsequent periods of sea-level rise, salinisation was apparently slow enough to allow relics of these freshwaters to be retained.

Onshore saline groundwater
In many flat coastal and delta areas the coastline during the recent geologic past was located further inland than today. As a result, vast quantities of saline water were retained in the subsurface after the sea level retreated. Such occurrences of saline groundwater are sometimes erroneously attributed to seawater intrusion, i.e. the inland movement of seawater due to aquifer over-exploitation (see Section 7.6.1). Clearly, effective water resource management requires proper understanding of the various forcing functions on the groundwater salinity distribution on a geological time-scale (Kooi and Groen, 2003).

High salinities are maintained for centuries to millennia, or sometimes even longer, when the presence of low-permeability deposits prevents flushing by meteoric water (e.g. Groen *et al.*, 2000; Yechieli *et al.*, 2001). Rapid salinisation due to convective sinking of seawater plumes occurs when the transgression is over a high-permeability substrate. This process is responsible for the occurrence of saline groundwater up to depths of 400 m in the coastal area of the Netherlands (Post and Kooi, 2003).

NUMERICAL MODELLING
In the analysis of transient effects of coastline migration on the salinity distribution in groundwater, variable density groundwater flow and solute transport modelling plays a vital role in developing comprehensive conceptual models.

The large spatial and time scales involved pose special challenges. In particular, the high-resolution model grid required to capture convective flow features imposes a severe computational burden that limits the size of the model domain. Other complications include the lack of information on past boundary conditions, insufficient data for proper parameterisation, especially for the offshore domain, and unresolved numerical issues with variable density codes. Mathematical modelling applications addressing offshore palaeo-groundwater have shown a typical progression in model complexity. Indeed, all modelling studies to date are cross-sectional only (2D). Earliest steady-state, sharp-interface, homogeneous permeability models (Meisler *et al.*, 1984) could not capture the marked disequilibrium conditions that are so apparent. The Meisler *et al.* (1984) model was only able to simulate the steady-state position of the (sharp) seawater–fresh groundwater interface. Different simulations were carried out for different sea-level low stands. The model was not capable of simulating the effect of the subsequent sea-level rise. With a non-steady, sharp-interface, single aquifer model, Essaid (1990) made a convincing case that the interface offshore Santa Cruz Country, California, has probably not achieved equilibrium with present-day sea-level conditions, but is still responding to the Holocene sea-level rise. Kooi *et al.* (2000) used a variable density flow model that included a moving coastline to study the transient behaviour of the salinity distribution in coastal areas during transgression (Figure 7.10). They found that Holocene transgression rates were often sufficiently high to cause the fresh–salt transition zone to lag behind coastline migration.

Person *et al.* (2003) and Marksammer *et al.* (2007) have conducted variable density modelling for the New England continental shelf, simulating sea-level change over several millions of years. Their results indicate that the first-order continental shelf topographic gradient is insufficient to cause freshening of aquifer systems to several hundreds of metres depth far out on the continental shelf. Continental ice-sheet recharge

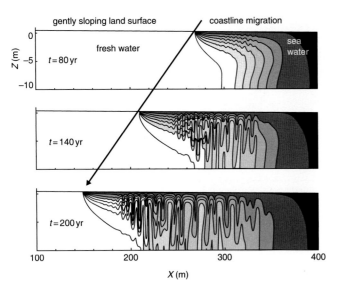

Figure 7.10. Variable density flow simulation showing salinisation lags behind coastline migration during transgression on a gently sloping surface and development of offshore brackish groundwater (after Kooi *et al.*, 2000). Highly unstable convective fingering is seen as dominant vertical salinisation mechanism.

(Marksammer *et al.*, 2007) and/or secondary flow systems associated with incised river systems in the continental shelf during low-stands therefore appear to be required to account for distal (several tens of kilometres off the coast, well beyond the reach of active topography-driven freshwater circulation) offshore palaeowaters.

FUTURE CHALLENGES
Knowledge of the 'anomalous' salinity distributions discussed above is not only relevant from an academic perspective. Offshore fresh and brackish groundwater occurrences are expected to become a viable alternative water resource, notably in arid climates and in places with high population densities. Moreover, in the development of conceptual hydrogeologic models of coastal aquifers, the relevance of transient phenomena should be considered more routinely than at present. Such an analysis has major implications for the numerical models based on these conceptualisations, such as the selection of appropriate boundary conditions and the choice of the time-scales that need to be considered.

The much-needed improvements in our understanding of the impact of coastline migration on hydrogeologic systems rely on more and better field observations. Age dating of groundwater, especially coastal groundwater, is extremely difficult but is an essential component of palaeohydrologic analyses. Exploration techniques needed to better delineate the subsurface salinity distribution require refinement and more versatility, especially in order to investigate offshore aquifers.

Finally, numerical models need to be improved as well, as has been outlined in Section 7.6.1. Difficulties currently arise in the accurate and physically realistic simulation of density-driven fingering due to grid discretisation requirements. Resolving the nature of the relationship between numerical features of a code and the physical fingering process is crucial since this unstable fingering process appears to be of prime importance in many hydrogeologic settings. Greater flexibility in boundary conditions are also required to allow for both heads and concentrations changes to occur due to changes in sea level. Land subsidence and sedimentation are among the additional complications that currently remain largely unexplored and poorly understood. The addition of these additional physical processes into a quantitative modelling framework will undoubtedly provide challenges for numerical models.

7.6.3 Aquifer storage and recovery

IMPORTANCE OF AQUIFER STORAGE
AND RECOVERY (ASR)
Aquifer storage and recovery (ASR) simply involves the injection and storage of freshwater in an aquifer. The theory was introduced by Cederstrom (1947) as an alternative to surface-water storage such as dams and reservoirs. The main advantages of ASR over surface-water storage include that it requires a relatively small area of land, potentially reduces the water lost to evaporation (e.g. from the surface of a dam), and to a certain extent the aquifer can behave as a water purification medium.

The original, idealised concept of ASR involved the injection of a cylindrical 'bubble' of freshwater out of a well, into a confined aquifer containing groundwater of a poorer quality (such that the groundwater itself would typically not be used). The bubble of water would then be stored in the aquifer until needed, at which point it would be pumped back out of the same well. In this situation, the volume of recoverable freshwater could be reduced by mixing between the injected and ambient waters, and by the injected plume migrating down-gradient (in the case where the water was injected into an aquifer where a background hydraulic gradient existed as is typical in most cases). In both cases, the recoverable volume will be smaller than the injected volume. Situations with minimal mixing and little migration down-gradient can allow almost 100% of the injected volume to be recovered. However, situations with large amounts of mixing and/or a significant migration of the plume down-gradient can lead to a very small (even zero) percentage of recoverable freshwater. In each specific situation, the percentage of recoverable freshwater (called the 'recovery efficiency') will determine whether an ASR operation will be economically viable, and whether it offers any advantages over surface-water storage technology.

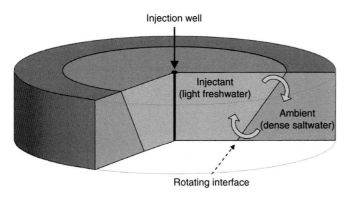

Figure 7.11. Basic concept of interface tilting leading to conical plume in the case of variable density analysis of aquifer storage and recovery.

VARIABLE DENSITY FLOW IN ASR

As well as mixing and down-gradient movement, density effects can cause a reduction in ASR recovery efficiency. It appears that this phenomenon was first documented by Esmail and Kimbler (1967). If freshwater is injected into a saltwater aquifer, the density contrast between injected and ambient water leads to an unstable interface at the edge of the injected plume. The denser ambient water tends to push towards the bottom of the interface and the lighter injected water tends to float towards the top. The result is tilting of the interface and a transition from the idealised cylindrical plume to a conical-shaped plume (see Figure 7.11). Upon recovery, the ambient water is much closer to the well at the bottom, and breakthrough of salt occurs, causing premature termination of the recovery before all of the freshwater has been extracted. In this way, it is possible to have situations with minimal mixing and minimal background hydraulic gradient, but still low recovery efficiency due to density effects.

ANALYTICAL MODELLING

Bear and Jacobs (1965) presented an analytical method for predicting the position of the freshwater–saltwater sharp interface under the influence of pumping (either injection or extraction) and a background hydraulic gradient. The analytical solution assumed a homogeneous aquifer and neglected density effects. Esmail and Kimbler (1967) produced an analytical–empirical formula to address the density-induced tilting of a freshwater–saltwater 'gradient' interface (i.e. accounting for a mixed zone) in radial geometry. The presence of a mixed zone was found to be significant. The influence of a background hydraulic gradient was neglected by Esmail and Kimbler (1967), and the process assumed that the interface would only tilt during storage, not during either the injection or extraction phase. Peters (1983) presented a semi-analytical solution for an injected bubble under the influence of density effects, but the presence of a mixed zone was neglected and the background hydraulic gradient was not considered. When considering the influence of pumping (especially multiple cycles), background hydraulic gradients,

heterogeneity and density-dependent flow phenomena, analytical solutions become far too complicated to derive and, as is often the case, a numerical model is required.

NUMERICAL MODELLING

Numerical modelling of ASR has been described by Merritt (1986), Missimer *et al.* (2002), Pavelic *et al.* (2006), Yobbi (1996) and Ward *et al.* (2007). Brown (2005) provides a comprehensive summary of ASR modelling (p. 170). Merritt (1986), Missimer *et al.* (2002) and Ward *et al.* (2007) specifically included results of numerical modelling with density effects explicitly accounted for. Yobbi (1996) included density effects in a modelling study but did not report on the specific influence and effects of density on the simulation results. Ward *et al.* (2007) performed numerical modelling of density-dependent ASR and quantified the influence of various parameters on the tilting interface.

Choice of model

An ASR process can be modelled in two dimensions as a horizontal plane (Merritt, 1986; Pavelic *et al.*, 2004). Being a radial flow system, it can also be modelled as a two-dimensional 'radial-symmetric' projection, i.e. a two-dimensional vertical projection that is symmetrical about the injection well (Merritt, 1986; Ward *et al.*, 2007). Of course, ASR can also be modelled in three dimensions but this obviously increases computational burden. Consequently, two-dimensional simplifications (in radial flow coordinates) are often preferred where symmetrical flow can be assumed.

Largely hypothetical ASR cases where there is no background hydraulic gradient, and where the aquifer structure is homogeneous or consists of infinite horizontal layers, can be modelled as a two-dimensional radial-symmetric projection (Ward *et al.*, 2007). Sample model output comparing density-invariant, small and large density contrasts are shown in Figure 7.12. These results were obtained using the FEFLOW model (Diersch, 2005). FEFLOW is a numerical finite element system that allows various two-dimensional projections as well as three-dimensional flow and transport processes in porous media. For this study a two-dimensional axisymmetric projection was chosen to model the ASR system. A transient, density-dependent flow and mass transport regime was chosen. Cases involving a background hydraulic gradient (more realistic) are inherently asymmetrical and must be modelled either as a two-dimensional horizontal or three-dimensional simulation.

To simulate ASR with variable density, flow must be considered in the vertical dimension, meaning that a two-dimensional horizontal plane model will not suffice. One concludes that density-dependent ASR systems must be modelled either as a two-dimensional radial-symmetric projection (which is incapable of simulating background hydraulic gradient or asymmetrical

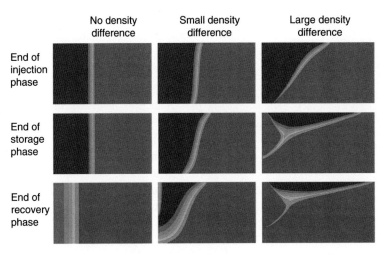

Figure 7.12 (Plate 3). An illustrative comparison between FEFLOW numerical model results for the cases of no density, small density and large density contrasts. Results are given for: end of injection phase, end of storage phase and end of recovery phase. Each image is symmetrical about its left-hand boundary, which represents the injection/extraction well where freshwater is injected into the initially more saline aquifer. The effects of variable density flow are clearly visible. Increasing the density contrast results in deviation from the standard cylindrical 'bubble' most, as seen in the no density difference case.

geological structures) or a fully three-dimensional model. Because in real systems a background hydraulic gradient is typical, a simple two-dimensional radial-symmetric projection is generally insufficient. However, the time and computational power required to establish and execute a three-dimensional model that is sufficiently discretised to accurately simulate the density-dependent flow of an injected freshwater plume can lead without robust quantification to the simpler choice of a density-invariant simulation model. This decision should not be made lightly, as judging the likely influence of density effects is rather complicated (Ward *et al.*, 2007).

When can density effects be ignored?

The concentration difference between injected and ambient water is the most obvious parameter driving the density-induced tilting of the interface. Indeed, in many cases the concentration difference is (explicitly or implicitly) assumed to be the only significant parameter that influences variable density flow, and modellers make the decision to use a density-dependent or density-invariant numerical model based on this parameter alone. However, Ward *et al.* (2007) demonstrated that an ASR system must be considered as a mixed-convective regime and derived the following mixed convection ratio M as an approximate ratio between the respective influences of free and forced convection in a simple (radial-symmetric) ASR system. In the case of an ASR system, they showed that the appropriate formulation of the mixed convection ratio (equation 7.2) is given by:

$$M = \frac{2\pi r B}{Q} K_V \left(\frac{\rho - \rho_0}{\rho_0} \right) \qquad (7.17)$$

where r is the hypothetical injected bubble radius (L), B is the aquifer thickness, Q is the injection or extraction flow rate (L^3/T), and K_V is the vertical hydraulic conductivity (L/T). When $M \ll 1$, forced convection dominates the system (e.g. fast pumping at small radii) and interface tilting is expected to be relatively insignificant. As M approaches 1, the relative influences of free and forced convection are comparable (e.g. slow pumping at large radii) and interface tilting is expected to become significant in this transition zone. Clearly during the storage phase, $Q = 0$ and M is infinite, indicating that the flow in system is totally dominated by free convection (Ward *et al.*, 2007). As can be seen from the mixed convection ratio given in equation (7.17), a low density contrast in a highly permeable medium may have the same 'density effect' (i.e. the same rate of tilting) as a higher density contrast in a less permeable medium. Equation (7.17) brings together several other potentially significant parameters, such as aquifer thickness, bubble radius and pumping rate. A highly anisotropic aquifer (in which the vertical hydraulic conductivity is a very small proportion of the horizontal) may retard the vertically acting density effects, thereby attenuating density-induced tilting. After injection finishes, the duration of the storage period is critical: depending on the density contrast and hydraulic conductivity, a short period of storage (e.g. <1 month) may lead to minimal tilting, but longer periods may cause significant tilting, and therefore significant reduction in recoverable freshwater. If ASR is to be promoted as a long-term drought mitigation strategy then long storage durations may be expected (Bloetscher and Muniz, 2004). In such cases, the effects of variable density flow are likely to be important in general hydrogeologic analyses and in numerical modelling efforts.

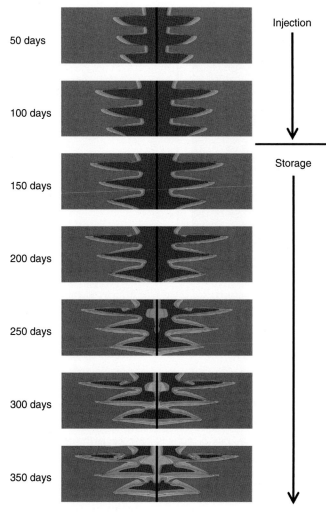

50 days

100 days

150 days

200 days

250 days

300 days

350 days

Injection

Storage

Figure 7.13 (Plate 4). Time series plot of a 100-day injection and 250-day storage, in an aquifer consisting of 6 even horizontal layers, with the hydraulic conductivity varying by a ratio of 10 between high and low K layers. Note the fingering processes.

Density effects in layered aquifers

Heterogeneous aquifers that consist of sedimentary strata can lead to density-induced fingering. Injected freshwater will readily flush ambient saltwater out of layers of high permeability, but saltwater will remain in layers of low permeability as the freshwater will not be able to penetrate those layers during injection. Then, during storage, the dense saltwater can migrate downwards out of the low-permeability layers and 'bleed' into the freshwater, and buoyant freshwater can also migrate upwards, flushing salt out of low-conductivity layers (see Figure 7.13). This can potentially lead to a very significant reduction in recovery efficiency as large volumes of freshwater stored in the high-permeability layers essentially become contaminated with salt from low-permeability layers above and below.

FUTURE CHALLENGES

When attempting to simulate ASR processes, one is faced with several problems. First, to model a realistic system with a background hydraulic gradient and density effects, a three-dimensional model is required. Setting up a detailed three-dimensional density-dependent model can be time-consuming and is generally computationally expensive. Furthermore, if the aquifer is heterogeneous, then the finite element mesh or grid may need to be highly discretised (perhaps in all three dimensions) at interfaces between regions of different hydraulic conductivity, and the run-times are likely to become excessively large (several days or more). The additional requirement to place model boundaries sufficiently far away from the injection/extraction activity will also increase simulation run-times. Large-field scale modelling applications of variable density flow in three dimensions are therefore limited because of heavy computational requirements.

Ward *et al.* (2007) described a method for simulating the injection/extraction well that considered the injection and extraction well to be a column of highly conductive elements (like a fracture), rather than a uniform flux across the boundary. The reason for this method of implementation was that the uniform flux method would fail to account for the dense water initially in the aquifer, which may in fact lead to a non-uniform velocity profile (low velocities at the bottom of the well and large velocities at the top of the well). Figures 7.12 and 7.13 demonstrate that the density-induced tilting of the interface can actually drive saltwater back up the well during the storage phase (when the well is not operational) where it can then circulate upwards into the upper part of the aquifer. The likelihood of observing this phenomenon in the field is unknown and requires further research.

7.6.4 Fractured rock flow and transport modelling

IMPORTANCE OF VARIABLE DENSITY FLOW IN FRACTURED ROCK AQUIFERS

Fractures may have a major impact on variable density flow because they represent preferential pathways where a dense plume may migrate at velocities that are several orders of magnitude faster than within the rock matrix itself. Studying dense plume migration in fractured rock is especially important in the context of hazardous waste disposal in low-permeability rock formations.

Variable density flow in a vertical fracture has previously been studied by Murphy (1979), Malkovsky and Pek (1997, 2004) and Shi (2005), showing that two-dimensional convective flow with a rotation axis normal to the fracture plane can occur.

Shikaze *et al.* (1998) numerically simulated variable density flow and transport in a network of discrete orthogonal fractures. They found that vertical fractures with an aperture as small as 50 μm significantly increase contaminant migration relative to

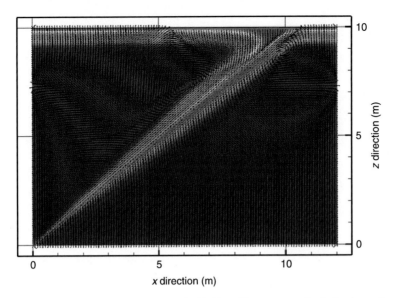

Figure 7.14 (Plate 5). Variable density flow in an inclined fracture embedded in a 2D porous matrix. Red colors refer to high concentration and blue colors refer to low concentration. Arrows represent groundwater flow velocities.

the case where fractures are absent. It was also shown that dense solute plumes develop in a highly irregular fashion and are extremely difficult to predict. However, Shikaze *et al.* (1998) represented fractures by one-dimensional segments, thus inhibiting convection within the fracture. Importantly, Shikaze *et al.* (1998) limited their studies to a regular fracture network consisting of only vertical and horizontal fractures, embedded in a porous matrix.

Graf and Therrien (2005) included inclined fractures in the simulation of dense plume migration and showed that variable density flow in a one-dimensional inclined fracture initiates convection in the two-dimensional porous matrix and that convection cells are independent and separated by the high-permeability fracture (Figure 7.14).

In a recent study, Graf and Therrien (2007a) investigated dense plume migration in irregular two-dimensional fracture networks. Simulations indicated that convection cells form and that they overlap both the porous matrix and fractures. Thus, transport rates in convection cells depend on matrix and fracture flow properties. A series of simulations in statistically equivalent networks of fractures with irregular orientation showed that the migration of a dense plume is highly sensitive to the geometry of the network. Simulations suggest that the fracture network dictates the number of fingers and that the number of fingers coincides with the number of equidistantly distributed fractures near the source (Graf and Therrien, 2007a). These fractures represent heterogeneities of the vertical conductivity and, therefore, disturb the flow field. Transient results showed that the disturbances grow in time to ultimately form instabilities. In conclusion, systems with low fracture density near the domain top but with at least one fracture near the source have the greatest potential to

be highly unstable. The number of instabilities in fractured media does not change with time.

Simmons *et al.* (1999) demonstrated that in a sandy aquifer the number of fingers decreases with time because large fingers increase, decreasing the number of small fingers. In fractured media, however, this is not the case because the location of fingers is strongly controlled by the geometry of the fracture network. In summary, if fractures in a random network are connected equidistantly to the solute source, few equidistantly distributed fractures favour density-driven transport while numerous fractures have a stabilising effect.

VERIFICATION OF VARIABLE DENSITY FLOW IN FRACTURED ROCK

Few test cases exist to verify a new code which simulates variable density flow in fractures. Results presented by Shikaze *et al.* (1998) can be used to verify dense plume migration in a set of vertical fractures by comparing isohalines at certain simulation times. The inclined-fracture problem introduced by Graf and Therrien (2005) can be used both qualitatively (using isohalines) and quantitatively (using detailed information on breakthrough curves, mass fluxes, maximum velocities, etc.) to verify variable density flow in an inclined fracture. Caltagirone (1982) has presented the analytical solution for the onset of convection in homogeneous media. The solution can be applied to vertical and inclined fractures by assuming constant fracture aperture (homogeneity) and by introducing a cosine weight to account for fracture incline (Graf, 2005).

Although the solution by Caltagirone (1982) is the only available analytical solution on variable density flow applied to box-type problems of finite domain dimensions, it is not truly an analytical solution, but rather uses analytical procedures to

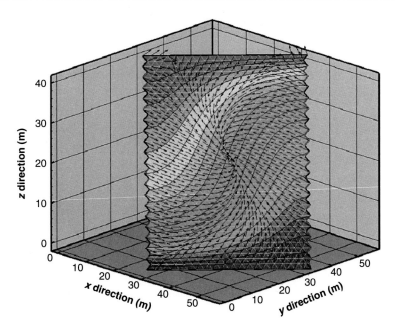

Figure 7.15 (Plate 6). Preliminary results of variable density flow simulations in an inclined non-planar fracture embedded in a 3D porous matrix. The fracture is discretised by 9 planar triangles +(triangulation) and the porous matrix cube is of side length 10 m.

derive the critical Rayleigh number for different aspect ratios in order to determine the conditions that govern the stable and unstable states of convection in box type problems. It is these solutions that also form the basis for newly proposed benchmark test cases presented in Weatherill *et al.* (2004). We conclude that, to date, a robust and well-accepted benchmark problem for variable density flow in fractures oriented in three dimensions and embedded in a 3D porous matrix does not exist. Thus, there is a clear need to further explore how to best test and benchmark 3D variable density flow in fractured media.

HYDROGEOSPHERE CODE CAPABILITY
The HydroGeoSphere (HGS) model simulates variable density flow in 3D porous media and in 2D fractured media where fractures are planar and parallel to at least one coordinate axis (Therrien and Sudicky, 1996; Therrien *et al.*, 2004; Graf and Therrien, 2005, 2007a, 2007b). There are, in principle, two methods to calculate density:

(i) Using Pitzer's ion interaction model in the VOPO model (Monnin, 1989, 1994), which is physically based but slow because it iterates between density and molality. The VOPO algorithm calculates the density of water from the concentrations of the solutes Na^+, K^+, Ca^{2+}, Mg^{2+}, Cl^-, SO_4^{2-}, HCO_3^- and CO_3^{2-}, based on Pitzer's ion interaction model. It allows for accurate calculation of the density of natural waters over a wide range of salinities (i.e. from freshwater to brine). However, the speed of the computation often makes it impractical for numerical modelling.

(ii) Using an approximate but faster empirical model.

HGS can use both methods and salinity is calculated from concentrations of individual ions found in natural water.

FUTURE CHALLENGES
Assuming 2D spatial dimensions is a common simplification made in numerical simulations. Because density-driven convection can naturally occur in multiple ways in three dimensions, representing non-planar fractures in three dimensions is essential. Prior attempts have shown that this is a tedious task but it has been addressed successfully (Graf and Therrien, 2006, 2007c). Variable density flow in inclined non-planar fractures embedded in a 3D porous matrix is a great future challenge and subject to ongoing investigation (Figure 7.15). Subsequently, benchmark problems for variable density flow in 3D fractured porous media are important and should be defined, followed by further 3D analyses.

7.6.5 Chemically reactive transport modelling

IMPORTANCE OF CHEMICAL REACTIONS IN VARIABLE DENSITY FLOW
Mixing of different waters, whether driven by the existence of salinity/temperature gradients or other forces, may trigger geochemical reactions. Well-documented saltwater-related problems include the dissolution of calcite in fresh–saltwater mixing zones (e.g. Wigley and Plummer, 1976; Sanford and Konikow, 1989) and cation exchange due to saltwater intrusion or freshening (Valocchi *et al.*, 1981; Appelo and Willemsen, 1987; Appelo *et al.*, 1990; Appelo, 1994). Precipitation of (carbonate) minerals

has also been extensively studied in geothermal problems due to its relevance for aquifer thermal energy storage (e.g. Brons *et al.*, 1991). Moreover, ore formation in relation to thermal convection is a topic that has received ample attention in basin-scale geological studies (e.g. Person *et al.*, 1996).

GOVERNING EQUATIONS

Reactive transport of dissolved solutes in groundwater is described by the advection–diffusion equation, which is a statement of the conservation of mass of a chemical component *i*:

$$\frac{\partial nC_i}{\partial t} = \nabla \cdot (nD \cdot \nabla C_i) - \nabla \cdot (qC_i) + r_{reac} \qquad (7.18)$$

where C_i denotes the chemical concentration of component *i* (M/L^3), *n* is the porosity (dimensionless), *D* is the dispersion coefficient (L^2/T) and *q* is the specific discharge (L/T), i.e. the volumetric flow rate per unit of cross-sectional area. The third term on the right-hand side, r_{reac}, accounts for the change in solute concentration due to chemical reactions. A similar expression holds for the temperature in heat-transport problems. Darcy's law in a variable density formulation is given in equation (7.13), and the general equation of state is given in equation (7.6). The concentration depends on the flow field (through *q* in equation 7.18), the flow field depends on the density (through ρ in equation 7.13) and ρ is influenced by *C* and hence also by *q*. Therefore, in variable density flow problems, the governing equations for groundwater flow and solute transport need to be solved simultaneously.

For reactive transport modelling, it is important to pay careful attention to the equation of state. This expression relates fluid density and dissolved solute concentration and can take different forms, for example a simple linear expression:

$$\rho = \rho_0 + \varepsilon C \qquad (19)$$

where ρ_0 is the density when $C = 0$, and ε is the slope of the ρ–C curve. In multicomponent systems, ρ is not a function of a single solute but depends on all the different solutes that are present in the solution. A more appropriate form of equation (7.19) then becomes:

$$\rho = \rho_0 + \sum_{i=1,n} \varepsilon_i C_i \qquad (20)$$

where *n* is the number of components and the subscripts *i* refer to the individual components. Relations of this type were employed by Zhang and Schwartz (1995) and Mao *et al.* (2006). Alternatively fluid density might be computed (slightly) more accurately on the base of a thermodynamic framework (Monnin, 1994; Freedman and Ibaraki, 2002; Post and Prommer, 2007).

DEVELOPMENT AND APPLICATIONS OF NUMERICAL MODELS

The development of numerical codes that address reactive transport under variable density conditions was driven by a range

of different motivations. On one hand, for example, large-scale transport phenomena over geological time-scales are simulated to understand the evolution and spatial distribution of ore-bodies (e.g. Raffensperger and Garven, 1995), while, on the other hand, models and simulations that focus on the movement of groundwater and the related water quality evolution are typically dealing with much smaller spatial and shorter time-scales. The latter, for example, was the case in the studies by Zhang and Schwartz (1995), who have developed a numerical simulator to assess the fate of contaminants in a dense leachate plume, and by Christensen *et al.* (2001), who studied hydrochemical processes during horizontal seawater intrusion. Furthermore, Mao *et al.* (2006) have developed a model and applied it to simulate the results of a tank simulation involving chemical reactions in an intruding seawater wedge. Sanford and Konikow (1989), Freedman and Ibaraki (2002) and Rezaeia *et al.* (2005) developed codes and used them for generic simulations that evaluated the effect of chemical reactions on dynamic changes of the hydraulic properties of the aquifer material.

Post and Prommer (2007) developed a variable density version of the reactive multicomponent transport model PHT3D (e.g. Prommer *et al.*, 2003; Prommer and Stuyfzand, 2005; Greskowiak *et al.*, 2005; Prommer *et al.*, 2006). Their model combines the previously existing and widely used tool SEAWAT, which accounts for multicomponent transport under variable density conditions, and the geochemical reaction simulator PHREEQC-2. They used the model to study the relative importance of reaction-induced density changes (resulting from cation exchange and calcite dissolution) on the groundwater flow field during free convection. To assess and quantify the potential effects, they reformulated the classic, well-studied Elder problem, in which flow is solely driven by density gradients (free convection), into a reactive multicomponent transport problem. The main outcome of the modelling study was that the flow field appears to be altered only in cases when the fluid density is strongly affected by changes in solute concentration due to chemical reactions (Figure 7.16). Perturbations smaller than approximately 10% did not result in differences in the flow pattern that could visually be detected, but an effect was noticed in the rate of salt plume descent that became higher as the density of the downward migrating plumes increased. The critical result here is that when low density-contrasts drive groundwater flow, chemical reactions are more likely to be important and hence need to be incorporated in the analysis, unlike when density contrasts are high, such as in seawater-intrusion type of problems.

The PHT3D code was also applied in the study of Bauer-Gottwein *et al.* (2007), simulating shallow, unconfined aquifers underlying islands in the Okavango delta (Botswana). These islands are fascinating natural hydrological systems, for which the interplay between variable density flow, evapotranspiration

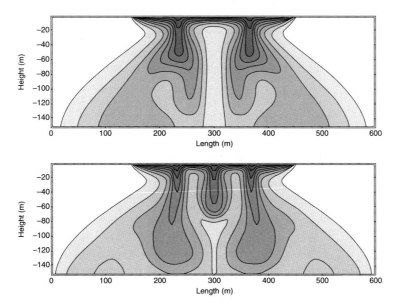

Figure 7.16. Chemically reactive adaptation of the Elder short heater natural convection problem. Width of the model domain is 600 m. Top: Plume configuration after 2920 days for the adapted Elder problem modelled by Post and Prommer [2007] when no chemical reactions are considered. Bottom: Plume configuration after 2920 days for the adapted Elder problem when cation exchange and calcite equilibrium are included in the simulation. Darker shading represents higher density.

and geochemical reactions has a crucial effect. Astonishingly, the combination of these mechanisms keeps the Okavango delta fresh, as salts are transported from the freshwater body towards the island's centre, where evapoconcentration triggers (i) mineral precipitation and (ii) convective, density-driven downward transport of the salt into deeper aquifer systems. The model simulations evaluated the conceptual model and quantified, using measured hydrochemical data as constraints, the individual processes and elucidated their relative importance for the system. They found that the onset of density-driven flow was affected by mineral precipitation and carbon-dioxide degassing as well as by complexation reactions between cations and humic acids.

FUTURE CHALLENGES
Numerical modelling of reactive transport under variable density conditions is a complex undertaking. As with transport under density-invariant conditions, numerical approximations introduce artificial dispersion and oscillations which pose special difficulties for chemical calculations. These are especially felt when redox problems are simulated as these require a high accuracy. The need therefore remains to search for mathematical techniques that combat these undesirable artefacts. In many cases, it should be noted that model results are extremely sensitive to spatial discretisation and that grid convergence can be a problem. There are also limitations faced by the assumption of local chemical equilibrium and how this relates to the choice of spatial and temporal discretisation (e.g. the time taken to equilibrium could be longer than the time-step size).

The capabilities of the codes developed so far are already quite impressive but existing limitations warrant further improvement and refinement. Multiphase systems, interaction with hydraulic properties and deforming porous media can already be handled by some codes but the development of more-comprehensive models is warranted due to the complexity of geological systems.

Despite these limitations, the level of sophistication of the reactive transport codes has already outrun our capability to collect the appropriate input data. Parameterisation of even the simplest models, such as well-controlled laboratory column experiments, is by no means trivial. For example, challenges relate to the translation of thermodynamic constants for ideal minerals to more amorphous phases in natural sediments or batch-determined kinetic laws to field conditions. Unless new techniques emerge (e.g. nanotechnology?) that allow us to change the way we collect field data, the reliable application of sophisticated codes will be seriously limited.

REFERENCES

Adams, J. J. and Bachu, S. (2002) *Equations of State for Basin Geofluids: Algorithm Review and Intercomparison for Brine.* Alberta Geological Survey, Edmonton, Canada.

Appelo, C. A. J. (1994) Cation and proton exchange, pH variations, and carbonate reactions in a freshening aquifer. *Water Resour. Res.* **30**(10), 2793–2805.

Appelo, C. A. J. and Willemsen, A. (1987) Geochemical calculations and observations on salt-water intrusions. 1. A combined geochemical mixing cell model. *Journal of Hydrology* **94**(3–4), 313–330.

Appelo, C. A. J., Willemsen, A., Beekman, H. E. and Griffioen, J. (1990) Geochemical calculations and observations on salt-water intrusions. 2. Validation of a geochemical model with laboratory experiments. *Journal of Hydrology* **120**(1–4), 225–250.

Badon-Ghyben, W. (1888) Nota in verband met de voorgenomen putboring nabij Amsterdam (Notes on the probable results of well drilling near Amsterdam). *Tijdschrift van het Koninklijk Instituut van Ingenieurs, The Hague,* **1888**/9, 8–22.

Barlow P. M. (2003) *Ground Water in Freshwater–Saltwater Environments of the Atlantic Coast.* Circular 1262. US Geological Survey, Reston.

Barth, G. R., Illangasekare, T. H., Hill, M. C. and Rajaram, H. (2001) A new tracer-density criterion for heterogeneous porous media, *Water Resour. Res.* **37**(1), 21–31.

Bauer-Gottwein, P., Langer, T., Prommer, H., Wolski, P. and Kinzelbach, W. (2007) Okavango Delta Islands: interaction between density-driven flow and geochemical reactions under evapo-concentration. *Journal of Hydrology* **335**, 389–405.

Bear, J. (1972) *Dynamics of Fluids in Porous Media.* American Elsevier.

Bear, J. (1979) *Hydraulics of Groundwater.* McGraw-Hill.

Bear, J. and Jacobs, M. (1965) On the movement of water bodies injected into aquifers. *Journal of Hydrology* **3**, 37–57.

Benard, H. (1900) Etude experimentale du mouvement des liquides propageuent de la chaleur par convection. Regime permament: tourbillons cellulaires. *C. R. Acad. Sci. Paris* **130**, 1004–1007.

Bloetscher, F. and Muniz, A. (2004) Can the results of modeling ASR systems answer long-term viability issues? *World Water and Environmental Resources Congress 2004,* Salt Lake City.

Boufadel, M. C., Suidan, M. T. and Venosa, A. D. (1997) Density-dependent flow in one-dimensional variably-saturated media, *J. Hydrol.* **202**(1–4), 280–301.

Brons, H. J., Griffioen, J., Appelo, C. A. J. and Zehnder, A. J. B. (1991) (Bio) geochemical reactions in aquifer material from a thermal energy storage site. *Water Resour. Res.* **25**(6), 729–736.

Brown, C. J. (2005) *Planning Decision Framework for Brackish Water Aquifer, Storage and Recovery (ASR) Projects.* University of Florida PhD Thesis.

Caltagirone, J. P. (1982) Convection in a porous medium. In Convective transport and instability phenomena, ed. J. Zierep and H. Oertel, 199–232 (chapter 1.3.2). Braunsche Hofbuchdruckerei und Verlag.

Cartwright, N., Li, L. and Nielsen, P. (2004) Response of the salt–freshwater interface in a coastal aquifer to a wave-induced groundwater pulse: field observations and modelling. *Advances in Water Resources* **27**, 297–303.

Cederstrom, D. J. (1947) Artificial recharge of a brackish water well. *The Commonwealth,* Dec. 1947, **31**.

Christensen, F. D., Engesgaard, P. and Kipp, K. (2001) A reactive transport investigation of a seawater intrusion experiment in a shallow aquifer, Skansehage Denmark. In *First International Conference on Saltwater Intrusion and Coastal Aquifers Monitoring, Modeling, and Management,* p. 4. Essaouira, Morocco.

Dagan G. and Zeitoun, D. G. (1998) Seawater–freshwater interface in a stratified aquifer of random permeability distribution. *Journal of Contaminant Hydrology* **29**, 185–203.

Diersch, H. J. G. (1988) Finite element modelling of recirculating density driven saltwater intrusion processes in groundwater. *Adv. Water Resources* **11**, 25–43.

Diersch, H. J. G. (2005) *FEFLOW Reference Manual.* WASY GmbH, Institute for Water Resources Planning and Systems Research, Berlin.

Diersch, H. J. G. and Kolditz, O. (2002) High-density flow and transport in porous media: approaches and challenges. *Adv. Water Resources,* **25**(8–12), 899–944.

Doherty, J. (2004) *Manual for PEST.* 5th edition. Watermark Numerical Computing, Australia. Available from http://www.sspa.com/pest. Cited 15 January 2005.

Duffy, C. J. and Al-Hassan, S. (1988) Groundwater circulation in a closed desert basin: Topographic scaling and climatic forcing. *Water Resour. Res.* **24**, 1675–1688.

Elder, J. W. (1967a) Steady free convection in a porous medium heated from below, *J. Fluid Mech.* **27**(1), 29–48.

Elder, J. W. (1967b) Transient convection in a porous medium, *J. Fluid Mech.* **27**(3), 609–623.

Esmail, O. J. and Kimbler, O. K. (1967) Investigation of the technical feasibility of storing fresh water in saline aquifers. *Water Resour. Res.* **3**(3), 683–695.

Essaid, H. I. (1990) A multilayerd sharp interface model of coupled freshwater and saltwater flow in coastal systems: model development and application. *Water Resour. Res.* **26**, 1431–1454.

FAO (1997) *Seawater Intrusion in Coastal Aquifers: Guidelines for Study, Monitoring and Control.* Water Reports 11, Food and Agriculture Organization of the United Nations, Rome.

Freedman, V. and Ibaraki, M. (2002) Effects of chemical reactions on density-dependent fluid flow: on the numerical formulation and the development of instabilities. *Adv. Water Resources* **25**(4), 439–453.

Frihy O. E., Debes, E. A. and El Sayed, W. R. (2003) Processes reshaping the Nile delta promontories of Egypt: pre- and post-protection. *Geomorphology* **53**(34), 263–279.

Frind, E. O. (1982a) Simulation of long term density-dependent transport in groundwater, *Adv. Water Resources* **5**, 73–88.

Frind, E. O. (1982b) Seawater intrusion in continuous coastal aquifer aquitard system, *Adv. Water Resources* **5**, 89–97.

Garven, G., Raffensperger, J. P., Dumoulin, J. A. *et al.* (2003) Coupled heat and fluid flow modeling of the Carboniferous Kuna Basin, Alaska: implications for the genesis of the Red Dog Pb-Zn-Ag-Ba ore district. *J. Geochem. Explor.* **78**–9, 215–219.

Gingerich, S. B. and Voss C. I. (2005) Three-dimensional variable-density flow simulation of a coastal aquifer in southern Oahu, Hawaii, USA. *Hydrogeology Journal* **13**, 436–450.

Glover, R. E. (1959) The pattern of fresh-water flow in a coastal aquifer, *J. Geophys. Res.,* **64**(4), 457–59.

Glover, R. E. (1964) The pattern of fresh-water flow in a coastal aquifer. In: Sea water in coastal aquifers. *US Geological Survey Water-Supply Paper* **1613**, 32–35.

Goswami, R. R. and Clement, T. P. (2007) Laboratory-scale investigation of saltwater intrusion dynamics. *Water Resour. Res.* **43**, Art. No. W04418.

Graf, T. (2005) Modeling coupled thermohaline flow and reactive solute transport in discretely-fractured porous media. PhD thesis, Université Laval, Québec, Canada. Available from http://www.theses.ulaval.ca.

Graf, T. and Therrien, R. (2005) Variable-density groundwater flow and solute transport in porous media containing nonuniform discrete fractures. *Advances in Water Resources* **28**(12), 1351–1367.

Graf, T. and Therrien, R. (2006) Discretizing non-planar discrete fractures for 3D numerical flow and transport simulations. *Annual Meeting of the Geological Society of America (GSA),* **38**(7), 462, Philadelphia PA/USA.

Graf, T. and Therrien, R. (2007a) Variable-density groundwater flow and solute transport in irregular 2D fracture networks. *Advances in Water Resources* **30**(3), 455–468.

Graf, T. and Therrien, R. (2007b) Coupled thermohaline groundwater flow and single-species reactive solute transport in fractured porous media. *Advances in Water Resources* **30**(4), 742–771.

Graf, T. and Therrien, R. (2007c) A method to discretize non-planar fractures for 3D subsurface flow and transport simulations. Manuscript submitted to *International Journal for Numerical Methods in Fluids.*

Greskowiak, J., Prommer, H., Vanderzalm, J., Pavelic, P. and Dillon, P. (2005) Modeling of carbon cycling and biogeochemical changes during injection and recovery of reclaimed water at Bolivar, South Australia. *Water Resour. Res.* **41**, W10418, doi: 10.1029/2005WR004095.

Groen, J. (2002) *The effects of transgressions and regressions on coastal and offshore groundwater,* PhD-thesis, Vrije Universiteit, Amsterdam, The Netherlands.

Groen, J., Velstra, J. and Meesters, A. G. C. A. (2000) Salinization processes in paleowaters in coastal sediments of Suriname: evidence from $\delta^{37}Cl$ analysis and diffusion modelling. *J. Hydrol.* **234**, 1–20.

Gvirtzman, H., Garven, G. and Gvirtzman, G. (1997) Hydrogeological modeling of the saline hot springs at the Sea of Galilee, Israel. *Water Resour. Res.* **33**(5), 913–926.

Gupta, N. and Bair, E. S. (1997) Variable-density flow in the midcontinent basins and arches region of the United States. *Water Resour. Res.* **33**(8), 1785–1802.

Gupta, S. K., Cole, C. R., Kincaid, C. T. and Monti, A. M. (1987) *Coupled Fluid, Energy and Solute Transport (CFEST) Model, Formulation and Users Manual,* BMI/ONWI-660.

Harbaugh, A. W. (2005) MODFLOW-2005: The US Geological Survey modular ground-water model – The ground-water flow process. *US Geological Survey Techniques and Methods* 6–A16.

Hassanizadeh, S. M. and Leijnse, T. (1988) On the modeling of brine transport in porous media. *Water Resour. Res.* **24**(2), 321–330.

Hathaway, J. C., Poag, C. W., Valentine, P. C. *et al.* (1979) US Geological Survey core drilling on the Atlantic shelf. *Science* **206**, 515–527.

Henry, H. R. (1959) Salt intrusion into freshwater aquifers. *J. Geophys. Res.* **64**, 1911–1919.

Henry, H. R. (1964) Interfaces between salt water and fresh water in coastal aquifers. *US Geol. Surv. Water Supply Paper* 1613-C, *Sea Water in Coastal Aquifers*, C35–C84.

Herbert, A. W., Jackson, C. P. and Lever, D. A. (1988) Coupled groundwater flow and solute transport with fluid density strongly dependent upon concentration. *Water Resour. Res.* **24**, 1781–1795.

Herzberg, A. (1901) Die Wasserversorgung einiger Nordseebder (The water supply of parts of the North Sea coast of Germany), *Z. Gasbeleucht. Wasserversorg.* **44**, 815–819.

Holzbecher, E. (1998) *Modeling Density-Driven Flow in Porous Media*, Springer.

Horton, C. W. and Rogers Jr., F. T. (1945) Convection currents in a porous medium. *J. App. Phys.* **16**, 367–370.

Huyakorn, P. S. and Taylor, C. (1976) Finite element models for coupled groundwater and convective dispersion, *Proc. 1st Int. Conf. Finite Elements in Water Resources*, 1.131–1.151. Pentech Press.

Huyakorn, P. S., Andersen, P. F., Mercer, J. W. and White, H. O. (1987) Saltwater intrusion in aquifers: development and testing of a three-dimensional finite element model. *Water Resour. Res.* **23**(2), 293–312.

HydroGeoLogic Inc. (2003) *MODHMS software (Version 2.0) documentation. I. Groundwater flow modules, II. Transport modules, III. Surface water flow modules*. HydroGeoLogic.

Ibaraki, M. (1998) A robust and efficient numerical model for analyses of density-dependent flow in porous media, *Jour. Of Contaminant Hydrology* **34**(3), 235–246.

Istok, J. D. and Humphrey, M. D. (1995) Laboratory investigation of buoyancy-induced flow (plume sinking) during 2-well tracer tests, *Ground Water* **33**(4), 597–604.

Jackson, C. P. and Watson, S. P. (2001) Modelling variable density groundwater flow. *Phys. Chem. Earth Pt B* **26**(4), 333–336.

Johannsen, K., Kinzelbach, W., Oswald, S. and Wittum, G. (2002) The saltpool benchmark problem – numerical simulation of saltwater upconing in a porous medium. *Advances in Water Resources* **25**(3), 335–348.

Johnson, T. A. and Whitaker, R. (2003) Saltwater intrusion in the coastal aquifers of Los Angeles County, California. In *Coastal Aquifer Management: Monitoring, Modeling, and Case Studies*, ed. A. H. D. Cheng and D. Ouazar, 29–48, Lewis Publishers.

Kacimov, A. R. and Sherif, M. M. (2006) Sharp interface, one-dimensional seawater intrusion into a confined aquifer with controlled pumping: analytical solution. *Water Resour. Res.* **42**, Art. No. W06501.

Kaleris, V., Lagas, G., Marczinek, S. and Piotrowski, J. A. (2002) Modelling submarine groundwater discharge: an example from the western Baltic Sea, *J. Hydrol.* **265**(1–4), 76–99.

Kim, Y., Lee, K. -S., Koh, D.-C. *et al.* (2003) Hydrogeochemical and isotopic evidence of groundwater salinisation in a coastal aquifer: a case study in Jeju volcanic island. *Korea. Journal of Hydrology* **270**, 282–294.

Kipp Jr, K. L. (1987) *HST3D: A computer code for simulation of heat and solute transport in three-dimensional ground-water flow systems*. US Geol. Survey Water-Res. Inv. Rept. 86–4095.

Koch, M. and Zhang, G. (1992) Numerical solution of the effects of variable density in a contaminant plume. *Groundwater* **30**(5), 731–742.

Kohout, F. A., Hathaway, J. C., Folger, D. W. *et al.* (1977) Fresh ground water stored in aquifers under the continental shelf: implications from a deep test, Nantucket Island, Massachusetts. *Wat. Res. Bull.* **13**, 373–386.

Kohout, F. A., Meisler, H., Meyer, F. W. *et al.* (1988) Hydrogeology of the Atlantic continental margin. In *The Geology of North America Vol. I-2 The Atlantic Continental Margin*, ed. R. E. Sheridan and J. A. Grow, 463–480. US Geol. Soc. Am.

Konikow, L. F., Sanford, W. E. and Campbell, P. J. (1997) Constant concentration boundary conditions: lessons from the HYDROCOIN variable-density groundwater benchmark problem. *Water Resour. Res.* **33**(10), 2253–2261.

Kontis, A. L. and Mandle, R. J. (1988) Modification of a three-dimensional ground-water flow model to account for variable density water density and effects of multiaquifer wells. *US Geol. Survey Water-Res. Inv. Rept.* 87–4265.

Kooi, H. and Groen, J. (2003) Geological processes and the management of groundwater resources in coastal areas. *Netherlands Journal of Geosciences*, **82**, 31–40.

Kooi, H., Groen, J. and Leijnse, A. (2000) Modes of seawater intrusion during transgressions, *Water Resour. Res.* **36**(12), 3581–3589.

Kretz, V., Berest, P., Hulin, J. P. and Salin, D. (2003) An experimental study of the effects of density and viscosity contrasts on macrodispersion in porous media. *Water Resour. Res.* **39**(2), Art. No. 1032.

Kuiper, L. K. (1983) A numerical procedure for the solution of the steady-state variable density groundwater flow equation. *Water Resour. Res.* **19**(1), 234–240.

Kuiper, L. K. (1985) Documentation of a numerical code for the simulation of variable density ground-water flow in three dimensions. *US Geol. Survey Water-Res. Inv. Rept.* 84–4302.

Langevin, C. D. (2003) Simulation of submarine ground water discharge to a marine estuary: Biscayne Bay, *Florida. Ground Water* **41**(6), 758–771.

Langevin, C. D. and Guo, W. (2006) MODFLOW/MT3DMS based simulation of variable density ground water flow and transport. *Ground Water* **44**(3), 339–351.

Langevin, C. D., Shoemaker, W. B. and Guo, W. (2003) MODFLOW-2000, the US Geological Survey Modular Ground-Water Model – Documentation of the SEAWAT-2000 Version with the Variable-Density Flow Process (VDF) and the Integrated MT3DMS Transport Process (IMT), *US Geological Survey Open-File Report* 03–426.

Lapwood, E. R. (1948) Convection of a fluid in a porous medium. *Proc. Cambridge Phil. Soc.* **44**, 508–521.

Le Blanc, D. R. (1984) Sewage plume in a sand and gravel aquifer, Cape Cod, Massachusetts. *US Geol. Surv. Water Supply Paper* 2218.

Le Blanc, D. R., Garabedian, S. P., Hess, K. M. *et al.* (1991) Large-scale natural gradient test in sand and gravel, Cape Cod, Massachusetts. 1. Experimental design and observed tracer movement. *Water Resour. Res.* **27**(5), 895–910.

Leijnse, A. and Hassanizadeh, S. M. (1989) *Verification of the METROPOL code for density dependent flow in porous media: HYDROCOIN Project, level 1, case 5, and level 3, case 4*, RIVM, Bilthoven, Netherlands, Rep 728528004.

Lemke, L. D., Abriola, L. M. and Goovaerts, P. (2004) Dense nonaqueous phase liquid (DNAPL) source zone characterization: Influence of hydraulic property correlation on predictions of DNAPL infiltration and entrapment, *Water Resour. Res.* **40**(1), Art. No. W01511.

Li, X. D. and Schwartz, F. W. (2004) DNAPL remediation with in situ chemical oxidation using potassium permanganate. I. Mineralogy of Mn oxide and its dissolution in organic acids. *J. Contam. Hydrol.* **68**(1–2), 39–53.

Lin, H. C., Richards, D. R., Yeh, G. T. *et al.* (1997) *FEMWATER: a three-dimensional finite element computer model for simulating density-dependent flow and transport in variably saturated media*. Technical Report CHL-97–12. Waterways Experiment Station, US Army Corps of Engineers, Vicksburg, MS 39180–6199.

Liu, H. H. and Dane, J. H. (1996) A criterion for gravitational instability in miscible dense plumes. *J. Contam. Hydrol.* **23**, 233–243.

Lusczynski, N. J. (1961) Head and flow of groundwater of variable density. *J. Geophy. Res.* **66**, 4247–4256

Ma, T. S., Sophocleous, M., Yu, Y-S. and Buddemeier, R. W. (1997) Modeling saltwater upconing in a freshwater aquifer in south central Kansas. *Journal of Hydrology* **201**, 120–137.

Malkovsky, V. I. and Pek, A. A. (1997) Conditions for the onset of thermal convection of a homogeneous fluid in a vertical fault. *Petrology* **5**(4), 381 387.

Malkovsky, V. I. and Pek, A. A. (2004) Onset of thermal convection of a single-phase fluid in an open vertical fault. *Physics of the Solid Earth* **40**(8), 672–679.

Mao, X., Prommer, H., Barry, D. A. *et al.* (2006) Three-dimensional model for multi-component reactive transport with variable density groundwater flow. *Environmental Modelling and Software* **21**, 615–628.

Marksammer, A. J., Person, M. A., Lewis, F. D. *et al.* (2007) Integrating geophysical, hydrochemical, and hydrologic data to understand the freshwater resources on Nantucket Island, Massachusetts. In *Data Integration in Subsurface Hydrology*, ed. D. W. Hyndman, F. D. Day-Lewis and K. Singha. AGU Water Resources Monograph.

Mazzia, A., Bergamaschi, L. and Putti, M. (2001) On the reliability of numerical solutions of brine transport in groundwater: analysis of infiltration from a salt lake. *Transp Porous Media* **43**(1), 65–86.

Mendoza, C. A. and Frind, E. O. (1990) Advective dispersive transport of dense organic vapours in the unsaturated zone. 1. Model development. *Water Resour. Res.* **26**(3), 379–387.

Meisler, H., Leahy, P. P. and Knobel, L. L. (1984) Effect of eustatic sea-level changes on saltwater-freshwater in the northern Atlantic coastal plain. USGS *Water Supply Paper* 2255.

Merritt, M. L. (1986) Recovering fresh water stored in saline limestone aquifers. *Ground Water* **24**(4), 516–529.

Missimer, T. M., Weixing Guo, Walker, C. W. and Maliva, R. G. (2002) Hydraulic and density considerations in the design of aquifer storage and recovery systems. *Florida Water Resources Journal*, Feb. 2002, 31–35.

Monnin, C. (1989) An ion interaction model for the volumetric properties of natural waters: density of the solution and partial molal volumes of electrolytes to high concentrations at 25°C. *Geochimica et Cosmochimica Acta* **53**, 1177–1188.

Monnin, C. (1994) Density calculation and concentration scale conversions for natural waters. *Computers and Geosciences* **20**(10), 1435–1445.

Murphy, H. D. (1979) Convective instabilities in vertical fractures and faults. *Journal of Geophysical Research* **84**(B11), 6121–6130.

Naji, A., Cheng, A. H. D., Ouazar, D. (1998) Analytical stochastic solutions of saltwater/freshwater interface in coastal aquifers. *Stochastic Hydrology and Hydraulics* **12**, 413–430.

Nield, D. A. (1968) Onset of thermohaline convection in a porous medium. *Water Resour. Res.* **11**, 553–560.

Nield, D. A. and Bejan, A. (1999) *Convection in Porous Media*, Springer-Verlag.

Nield, D. A. and Bejan, A. (2006) *Convection in Porous Media*, 3rd edn. Springer.

Nield, D. A. and Simmons, C. T. (2006) A discussion on the effect of heterogeneity on the onset of convection in a porous medium, *Transport in Porous Media*, DOI 10.1007/s11242–006–9045–8.

Nihoul, J. C. J., Zavialov, P. O. and Micklin, P. P. (2004) Dying and dead seas climatic versus anthropic causes. *NATO Science Series IV: Earth and Environmental Sciences*, Kluwer Academic.

Oldenburg, C. M. and Pruess, K. (1995) Dispersive transport dynamics in a strongly coupled groundwater-brine flow system. *Water Resour. Res.* **31**(2), 289–302.

Oldenburg, C. M. and Pruess, K. (1999) Plume separation by transient thermohaline convection in porous media. *Geophys. Res. Lett.* **26**(19), 2997–3000.

Oostrom, M., Dane, J. H., Güven, O. and Hayworth, J. S. (1992) Experimental investigation of dense solute plumes in an unconfined aquifer model. *Water Resour. Res.* **28**(9), 2315–2326.

Oostrom, M., Thorne, P. D., White, M. D., Truex, M. J. and Wietsma, T. W. (2003) Numerical modeling to assess DNAPL movement and removal at the scenic site operable unit near Baton Rouge, Louisiana: a case study. *Soil Sediment Contam.* **12**(6), 901–926.

Ophori, D. U. (1998) The significance of viscosity in density-dependent flow of groundwater. *J. Hydrol.* **204**(1–4), 261–270.

Ouyang, Y. and Zheng, C. (1999) Density-driven transport of dissolved chemicals through unsaturated soil. *Soil Sci.* **164**(6), 376–390.

Paschke, N. W. and Hoopes, J. A. (1984) Buoyant contaminant plumes in groundwater, *Water Resources Res.* **20**(9), 1183–1192.

Pavelic, P., Dillon, P. and Robinson, N. (2004) *Groundwater Modelling to Assist Well-Field Design and Operation for the ASTR Trial at Salisbury, South Australia*. CSIRO Land and Water Technical Report No. 27/04, Commonwealth Science and Industrial Research Organisation (CSIRO), Adelaide, Australia.

Pavelic, P., Dillon, P. J. and Simmons, C. T. (2006) Multi-scale characterization of a heterogeneous aquifer using an ASR operation, *Ground Water* **44**(2), 155–164.

Person, M., Raffensperger, J. P., Ge, S. and Garven, G. (1996) Basin-scale hydrogeologic modeling. *Reviews of Geophysics* **34**(2), 307–309.

Person, M., Dugan, B., Swenson, J. B. *et al.* (2003) Pleistocene hydrogeology of the Atlantic continental shelf, New England. *Geol. Soc. Am. Bull.* **115**, 1324–1343.

Peters, J. H. (1983) The movement of fresh water injected in salaquifers, *Geologia Applicata e Idrogeologia* **18**(2), 144–155.

Pinder, G. F. and Cooper, H. H. (1970) A numerical technique for calculating the transient position of the saltwater front, *Water Resour. Res.* **6**(3), 875–882.

Post, V. E. A. and Kooi, H. (2003) On rates of salinization by free convection in high-permeability sediments: insights from numerical modelling and application to the Dutch coastal area. *Hydrogeology Journal* **11**, 549–559.

Post, V. E. A. and Prommer, H. (2007) Multicomponent reactive transport simulation of the Elder problem: effects of chemical reactions on salt plume development. *Water Resour. Res.* **43**(10), art. no. W10404.

Post, V. E. A., Hooijboer, A., Groen, J., Gieske, J. and Kooi, H. (2000) Pore water chemistry of clay layers in the southern North Sea: clues to the hydrogeological evolution of coastal areas. In *Proceedings 16th SWIM Conference, Miȩdzyzdroje, Poland*, 12–15 June 2000, 127–132.

Post, V. E. A., Kooi, H. and Simmons, C. T. (2007) Using hydraulic head measurements in variable-density ground water flow analyses. *Ground Water* **45**(6), 664–671.

Prommer, P. and Stuyfzand, P. (2005) Identification of temperature-dependent water quality changes during a deep well injection experiment in a pyritic aquifer. *Environmental Science and Technology* **39**, 2200–2209.

Prommer, H., Barry, D. A. and Zheng, C. (2003) MODFLOW/MT3DMS-based reactive multicomponent transport modelling. *Ground Water* **42**(2), 247–257.

Prommer, H., Tuxen, N. and Bjerg, P. L. (2006) Fringe-controlled natural attenuation of phenoxy acids in a landfill plume: integration of field-scale processes by reactive-transport modelling. *Environ. Sci. Technol.* **40**, 4732–4738.

Raffensperger, J. P. and Garven, G. (1995) The formation of unconformity-type uranium ore deposits. 2. Coupled hydrochemical modeling. *American Journal of Science* **295**, 639–696.

Raffensperger, J. P. and Vlassopoulos, D. (1999) The potential for free and mixed convection in sedimentary basins. *Hydrogeology J.* **7**, 505–520.

Rayleigh, Lord (J. W. Strutt) (1916) On convection currents in a horizontal layer of fluid when the higher temperature is on the under side. *Phil. Mag.* **6**(32), 529–546.

Reeves, M., Ward, D. S., Johns, N. D. and Cranwell, R. M. (1986) *Theory and Implementation for SWIFT II, the Sandia Waste Isolation Flow and Transport Model for Fractured Media, Release 4.84*. NUREG/CR-3328, SAND83–1159, Sandia National Laboratories, Albuquerque, NM, USA.

Rezaeia, M., Sanz, E., Raeisia, E. *et al.* (2005) Reactive transport modeling of calcite dissolution in the fresh-salt water mixing zone. *Journal of Hydrology* **311**, 282–298.

Robinson, C., Gibbes, B. and Li, L. (2006) Driving mechanisms for flow and salt transport in a subterranean estuary. *Geophysical Research Letters*, **33**, L03402. doi: 10.1029/2005GL025247.

Robinson, C., Li, L. and Barry, D. A. (2007a) Effect of tidal forcing on a subterranean estuary. *Advances in Water Resources* **30**, 851–865.

Robinson, C., Gibbes, B., Carey, H. and Li, L. (2007b) Salt-freshwater dynamics in a subterranean estuary over a spring-neap tidal cycle, *Journal of Geophysical Research*.

Robinson, C., Li, L. and Prommer, H. (2007c) Tide-induced recirculation across the aquifer-ocean interface, *Water Resour. Res.*

Sanford, W. E. and Konikow, L. F. (1985) A two-constituent solute-transport model for groundwater having variable density. *US Geol. Surv. Water Resour. Invest. Rep.* 85–4279.

Sanford, W. E. and Konikow, L. F. (1989) Simulation of calcite dissolution and porosity changes in saltwater mixing zones in coastal aquifers. *Water Resour. Res.* **25**(4), 655–667.

Sanford, W. E. and Wood, W. W. (2001) Hydrology of the coastal sabkhas of Abu Dhabi. *United Arab Emirates Hydrogeol J.* **9**(4), 358–366.

Schincariol, R. A. and Schwartz, F. W. (1990) An experimental investigation of variable density flow and mixing in homogeneous and heterogeneous media. *Water Resour. Res.* **26**(10), 2317–2329.

Schincariol, R. A., Schwartz, F. W. and Mendoza, C. A. (1997) Instabilities in variable density flows: stability analyses for homogeneous and heterogeneous media. *Water Resour. Res.* **33**(1), 31–41.

Schotting, R. J., Moser, H. and Hassanizadeh, S. M. (1999) High-concentration-gradient dispersion in porous media: experiments, analysis and approximations. *Adv. Water Resources* **22**(7), 665–680.

Segol, G. (1993) *Classic Groundwater Simulations: Proving and Improving Numerical Models*. Prentice Hall.

Segol, G. and Pinder, G. F. (1976) Transient simulation of salt water intrusion in south eastern Florida. *Water Resour. Res.* **12**, 65–70.

Senger, R. K. (1993) Paleohydrology of variable-density groundwater-flow systems in mature sedimentary basins – example of the Palo Duro Basin, Texas, USA. *J. Hydrol.* **151**(2–4), 109–145.

Sharp Jr., J. M., Fenstemaker, T. R., Simmons, C. T., McKenna, T. E. and Dickinson, J. K. (2001) Potential salinity-driven free convection in a shale-rich sedimentary basin: Example from the Gulf of Mexico basin in south Texas. *AAPG Bulletin*, **85**(12) 2089–2110.

Sherif, M. (1999) Nile delta aquifer in Egypt. In *Seawater Intrusion in Coastal Aquifers*, ed. J. Bear *et al.*, 559–590. Kluwer Academic.

Shi, M. (2005) *Characterizing heterogeneity in low-permeability strata and its control on fluid flow and solute transport by thermalhaline free convection.* Unpublished PhD thesis, University of Texas at Austin; 229 pp.

Shibuo, Y., Jarsjo, J. and Destouni, G. (2006) Bathymetry-topography effects on saltwater-fresh groundwater interactions around the shrinking Aral Sea. *Water Resour. Res.* **42**, art. no. W11410.

Shikaze, S. G., Sudicky, E. A. and Schwartz, F. W. (1998) Density-dependent solute transport in discretely-fractured geologic media: is prediction possible? *Journal of Contaminant Hydrology* **34**(10), 273–291.

Simmons, C. T. (2005) Variable density groundwater flow: from current challenges to future possibilities. *Hydrogeology Journal* **13**(1), 116–119.

Simmons, C. T., Narayan, K. A. and Wooding, R. A. (1999) On a test case for density-dependent groundwater flow and solute transport models: the salt lake problem. *Water Resour. Res.* **35**(12), 3607–3620.

Simmons, C. T., Fenstemaker, T. R. and Sharp Jr., J. M. (2001) Variable-density groundwater flow and solute transport in heterogeneous porous media: Approaches, resolutions and future challenges. *J. Contam. Hydrol.* **52**(1–4), 245–275.

Simmons, C. T., Narayan, K. A., Woods, J. A. and Herczeg, A. L. (2002) Groundwater flow and solute transport at the Mourquong saline-water disposal basin, Murray Basin, southeastern Australia, *Hydrogeol J.* **10**(2), 278–295.

Therrien, R. and Sudicky. E. A. (1996) Three-dimensional analysis of variably-saturated flow and solute transport in discretely-fractured porous media. *Journal of Contaminant Hydrology* **23**, 1–44.

Therrien, R., McLaren, R. G., Sudicky, E. A. and Panday, S. M. (2004) *HYDROGEOSPHERE– A three-dimensional numerical model describing fully-integrated subsurface and surface flow and solute transport.* Université Laval, University of Waterloo.

Thorenz, C., Kosakowski, G., Kolditz, O. and Berkowitz, B. (2002) An experimental and numerical investigation of saltwater movement in a coupled saturated-partially saturated systems. *Water Resour. Res.* **38**(6), Art. No. 1069.

Underwood, M. R., Peterson, F. L. and Voss, C. I. (1992) Groundwater lens dynamics of atoll islands. *Water Resour. Res.* **28**(11), 2889–2902.

Valocchi, A. J., Street, R. L. and Roberts, P. V. (1981) Transport of ion-exchanging solutes in groundwater – chromatographic theory and field simulation. *Water Resour. Res.* **17**(5), 1517–1527.

van Dam, J. C. (1999) Exploitation, restoration and management. In *Seawater Intrusion in Coastal Aquifers*, ed. J. Bear *et al.*, 73–125. Kluwer Academic.

Voss, C. I. (1984) SUTRA: a finite-element simulation model for saturated-unsaturated fluid density-dependent groundwater flow with energy transport or chemically reactive single-species solute transport. *US Geol. Surv. Water Resources Invest. Rep.* 84–4369.

Voss, C. I. (1999) USGS SUTRA code – history, practical use, and application in Hawaii. In *Seawater Intrusion in Coastal Aquifers*, ed. J. Bear *et al.*, 249–313, Kluwer Academic.

Voss, C. I. and Souza, W. R. (1987) Variable density flow and solute transport simulation of regional aquifers containing a narrow freshwater-saltwater transition zone. *Water Resources Res.* **23**(10), 1851–1866.

Voss, C. I., Provost, A. M. and SUTRA (2003) A model for saturated-unsaturated, variable-density ground-water flow with solute or energy transport. *US Geological Survey Water-Resources Investigations Report 02-4231.* US Geological Survey.

Voss, C. I., Simmons, C. T. and Robinson, N. I. (2010) Three dimensional benchmark for variable-density flow and transport simulation: matching

semi-analytic stability modes for steady unstable convection in an inclined porous box. *Hydrogeology Journal* **18**(1), 5–23.

Ward, J., Simmons, C. T. and Dillon, P. J. (2007) A theoretical analysis of mixed convection in aquifer storage and recovery: how important are density effects? *Journal of Hydrology* **343**(3), 169–186.

Weatherill, D., Simmons, C. T., Voss, C. I. and Robinson, N. I. (2004) Testing density-dependent groundwater models: two-dimensional steady state unstable convection in infinite, finite and inclined porous layers. *Advances in Water Resources* **27**, 547–562.

Welty, C., Kane A. C. and Kauffman, L. J. (2003) Stochastic analysis of transverse dispersion in density-coupled transport in aquifers. *Water Resour. Res.* **39**(6), Art. No. 1150.

Werner, A. D. and Gallagher, M. R. (2006) Characterisation of sea-water intrusion in the Pioneer Valley, Australia using hydrochemistry and three-dimensional numerical modelling. *Hydrogeology Journal* **14**, 1452–1469.

Werner, A. D. and Lockington, D. A. (2006) Tidal impacts on riparian salinities near estuaries. *Journal of Hydrology* **328**, 511–522.

Wigley, T. M. L. and Plummer, L. N. (1976) Mixing of carbonate waters. *Geochimica et Cosmochimica Acta* **40**(9), 989–995.

Williams, M. D. and Ranganathan, V. (1994) Ephemeral thermal and solute plumes formed by upwelling groundwaters near salt domes, *J. Geophys. Research-Solid Earth* **99**(b8), 15667–15681.

Wood, J. R. and Hewett, T. A. (1984) Reservoir diagenesis and convective fluid flow. In *Clastic Diagenesis*, ed. D. A. McDonald and R. C. Surdam. *Am. Assoc. Petroleum Geologists Memoir* **37**, 99–111.

Wooding, R. A. (1959) The stability of a viscous liquid in a vertical tube containing porous material. *Proc. Roy. Soc. London, Ser. A* **252**, 120–134.

Wooding, R. A. (1962) The stability of an interface between miscible fluids in a porous medium. *Z. Agnew. Math. Phys.* **13**, 255–265.

Wooding, R. A. (1963) Convection in a saturated porous medium at large Rayleigh number or Peclet number. *J. Fluid Mech.* **15**, 527–544.

Wooding, R. A. (1969) Growth of fingers at an unstable diffusing interface in a porous medium or Hele-Shaw cell. *J. Fluid Mech.* **39**, 477–495.

Wooding, R. A., Tyler, S. W. and White, I. (1997a) Convection in groundwater below an evaporating salt lake. 1. Onset of instability. *Water Resour. Res.* **33**, 1199–1217.

Wooding, R. A., Tyler, S. W., White, I. and Anderson, P. A. (1997b) Convection in groundwater below an evaporating salt lake. 2. Evolution of fingers or plumes. *Water Resour. Res.* **33**, 1219–1228.

Yechieli, Y. and Wood, W. W. (2002) Hydrologic processes in saline systems: playas, sabkhas, and saline lakes. *Earth-Sci. Rev.* **58**(3–4), 343–365.

Yechieli, Y., Kafri, U., Goldman, M. and Voss, C. I. (2001) Factors controlling the configuration of the fresh-saline water interface in the Dead Sea coastal aquifers: synthesis of TDEM surveys and numerical groundwater modeling. *Hydrogeology Journal* **9**(4), 367–377.

Ying, O. Y. and Zheng, C. M. (1999) Density-driven transport of dissolved chemicals through unsaturated soil. *Soil Sci.* **164**(6), 376–390.

Yobbi, D. K. (1996) *Simulation of Subsurface Storage and Recovery of Treated Effluent Injected in a Saline Aquifer, St Petersburg, Florida.* Water-Resources Investigations Report, US Geological Survey.

Zhang, H. B. and Schwartz, F. W. (1995) Multispecies contaminant plumes in variable-density flow systems. *Water Resour. Res.* **31**(4), 837–847.

Zhang, H. B., Schwartz, F. W., Wood, W. W., Garabedian, S. P. and LeBlanc, D. R. (1998) Simulation of variable-density flow and transport of reactive and nonreactive solutes during a tracer test at Cape Cod, Massachusetts. *Water Resour. Res.* **34**(1), 67–82.

8 Sustainable water management in arid and semi-arid regions

W. Kinzelbach, P. Brunner, A. von Boetticher, L. Kgotlhang and C. Milzow

8.1 INTRODUCTION

As a result of the ever-growing global population, pressure on water resources is increasing continuously, above all in arid and semi-arid regions. In many cases, the presently applied management practices are non-sustainable and lead to serious water-related problems such as the depletion of aquifers, the accumulation of substances to harmful levels, to water conflicts or economically infeasible costs. In this contribution, we illustrate these problems with several case studies. The northwest Sahara aquifer system is used to show the consequences of the overpumping of aquifers. A typical upstream–downstream problem is discussed with the example of the Okavango delta. Another case study discussed is the Yanqi basin in China. The Yanqi basin is a typical example showing how inappropriate irrigation practices can lead to soil salinisation and ecological problems of the downstream. Possible solutions to these problems and the role of numerical modelling as a tool to develop sustainable management practices are discussed. Some of the most common problems in setting up reliable models are highlighted and ideas how to address these problems are given. In the last part of this chapter, we discuss strategies to close the existing gap in thinking between scientists and decision makers.

8.2 DEFINITION OF SUSTAINABLE WATER MANAGEMENT

In a pragmatic definition, sustainable water management describes a practice which prevents irreversible damage to the resource water and resources related to it such as soils and ecosystems and which preserves in the long-term the ability of the resource to extend its services (including ecological services). More ambitious approaches include the triad of environmental conservation, economic efficiency and social equity under the heading of sustainability. While this concept is broad enough to address a wide range of systems it is not specific enough to be of help in a concrete problem. Considering this trade-off, it is easier to define non-sustainable practices that should be avoided. Such a non-sustainable practice is a practice which is hard to change and yet cannot go on indefinitely without running into a crisis. This type of non-sustainability shows:

- in the depletion of a finite resource, which cannot be substituted – examples include overpumping of groundwater resources in arid regions, degradation of soils, destruction of ecosystems or loss of biodiversity;
- in the accumulation of substances to harmful levels – such substances can include nutrients, salts and heavy metals, for example;
- in the unfair allocation of a resource leading to conflict – for example, upstream–downstream problems along a river or problems arising from the simultaneous abstraction by a number of users from a common groundwater resource;
- in runaway cost of water.

In the definition of sustainability the choice of the system boundary both in space and time is crucial. While a practice may guarantee local sustainability it might jeopardise the downstream sustainability. A practice which does not show serious consequences at present may well do so in the future. Similarly, it makes no sense to distinguish between surface and groundwater management. The two compartments are interrelated. Groundwater is not an independent water resource. Rivers can drain or recharge aquifers. Overall, groundwater will eventually surface in springs, as base flow of rivers, discharge to the sea, or as evapotranspiration. An aquifer basically constitutes a storage device, which delays the runoff of surface water.

In the following we concentrate on the pragmatic definition of sustainability. We discuss problems of sustainability and illustrate them with examples from our own research. They concern:

- the overpumping of aquifers, illustrated by the northwest Sahara aquifer system;
- the upstream–downstream conflict and associated threats to ecosystems, illustrated by the Okavango delta;

Groundwater Modelling in Arid and Semi-Arid Areas, ed. Howard S. Wheater, Simon A. Mathias and Xin Li. Published by Cambridge University Press.

- the salinisation of soils due to inappropriate irrigation practices, illustrated by the Yanqi basin, China;
- the pollution of water bodies with large residence time by persistent or recyclable pollutants, illustrated by seawater intrusion.

8.3 OVERPUMPING OF AQUIFERS

8.3.1 General situation

Groundwater contributes worldwide with about 20% to the freshwater supply. Despite this relatively small proportion its role is important for two reasons: On the one hand, groundwater is well suited for the supply of drinking water due to its usually high quality. On the other hand, groundwater basins are important long-term storage reservoirs, which in semi-arid and arid countries often constitute the only perennial water resource. The storage capacity is evident if one compares the volumes of surface and groundwater resources. Globally the volume of freshwater resources in rivers and lakes is about $100\,000$ km^3. With about $10\,000\,000$ km^3, the volume of groundwater is two orders of magnitude larger (e.g. Gleick, 1993). For a sustainable water management, however, the renewal rate is more relevant and for this quantity the situation is just reversed. The renewal rate of surface-water resources is $30\,000$ km^3 per year, that of groundwater only about 3000 km^3 per year. Worldwide, about 800 km^3 of groundwater are utilised by humans annually. This number still looks considerably smaller than the yearly renewal rate. However, the global comparison does not do justice to the real situation. Averaged figures hide the fact that about one quarter of the yearly withdrawal rate is supplied by non-renewable fossil groundwater reserves (Sahagian *et al.*, 1994).

Groundwater plays a vital role for agricultural and domestic water supply. On a global scale 20% of the arable land is irrigated (FAO, 2006) out of which a large portion is entirely based on groundwater. The main reason for groundwater depletion is often large-scale, groundwater-based irrigation. The most extreme cases are India, Bangladesh and North China, where groundwater depletion by agriculture has reached critical levels.

The expansion of irrigated areas can significantly influence the local supply and demand balance of water. If the gap between limited renewable water sources and a continuously increasing water demand is compensated by groundwater provided from storage, the groundwater table inevitably drops. The drawdown propagates towards the boundaries of the aquifer, diminishing or completely terminating the drainage flows. If the drawdown cones become deep, the remaining water resources are more difficult to access. In areas where users were accustomed to a large and inexpensive supply, agricultural production will break

down as the available resources become scarce (Postel, 1997). On the Arabian peninsula, in North Africa, China and the arid Western United States for example, abstractions for large-scale irrigation have withdrawn large quantities of fossil water, which under present climatic conditions are no longer replenished. In the Ogallala aquifer large drawdowns have virtually extinguished the former irrigated maize culture due to high water cost. In the North China plain, irrigated wheat cultivation is being abandoned due to high water cost and is moving to the wetter northeast, where wheat can be grown by rain-fed agriculture.

8.3.2 Example: The northwest Sahara aquifer system

In northern Africa, fossil groundwater from the North West Sahara Aquifer System (NWSAS) is being mined for the provision of irrigation water. This groundwater system consists of two main water-bearing structures, the Terminal Complex (TC) and the underlying Intercalary Continental (IC). Algeria, Tunisia and Libya together share this transboundary resource, which is one of the largest groundwater systems on the planet (see Figure 8.1). While there is enough freshwater to last for many centuries at present abstraction rates, it comes at an increasing cost. Resource exploitation generally causes a decrease of the ratio of resource to capital employed due to the general lowering of the piezometric heads and a deterioration of groundwater quality. The problem of the three countries' water authorities is to allocate water in time and space adequately in order to avoid excessive allocation costs (Observatoire du Sahara et du Sahel (OSS), 2003).

The NWSAS is a common property resource that exhibits stock externalities. In the highly exploited central region of the basin for example, declining piezometry is a composite effect caused by local pumping and regional transboundary abstraction. Therefore, the allocation strategies of the individual countries are conflicting. Between individually cost-effective policies on a national level, many trade-off solutions to the above stated allocation problem exist. In fact, each spatial configuration of resource allocation is a trade-off between the individual objectives of the planners involved. While a deepening cone of depression increases pumping cost, a spreading out of wells leads to higher transmission costs. Optimisation eventually implies a degree of cooperation possibly involving transboundary water transfers (Siegfried and Kinzelbach, 2006).

An agreement on coordinated, cooperative allocation is only possible if it implies an improvement over the non-cooperative allocation for each decision maker. Instead of a single optimal allocation policy, a decision-supporting management tool should identify a set of compromise solutions, for which none of the trade-offs contained can be said to be better than the others in the absence of preference information (Binmore, 1994). Such a set of policies is generally referred to as Pareto-optimal.

Table 8.1. *Mean unit groundwater allocation prices per cubic meter water applied to irrigation in the years 2000 and 2050. In the context of the NWSAS, prices are in relative units because absolute values were difficult to establish. Their ratios, however, remain valid.*

	Non-cooperative		Cooperative	
	2000	2050	2000	2050
Algeria	0.031	2.8	0.021	0.65
Tunisia	0.057	1.8	0.006	0.18
Libya	0.065	1.9	0.019	0.28

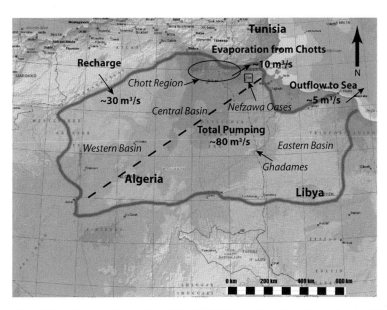

Figure 8.1 (Plate 7). Map of northern Africa showing the outline of the NWSAS as well as the estimated water balance for the year 2000 (Siegfried, 2004).

Allocation costs over a certain time horizon can be calculated by coupling a groundwater model and an economic model. The latter calculates present costs, consisting of energy costs (pumping and conveyance requirements) and installation costs of allocation policies, at a given future water demand, factor prices and discount rates. Economic scarcity can be taken into consideration by limiting feasible drawdowns to a certain depth from where on pumping becomes too expensive with respect to agricultural production cost. Imposed gradient constraints at locations that are sensitive to groundwater salinisation ensure the latter not to occur (Siegfried, 2004).

Based on demand and price predictions, a model was run for the time from 2000 to 2050. Model results show that the average per unit water provision prices in the year 2050 are 5 times smaller for an adaptive strategy relocating well fields, compared to a strategy which keeps the present pumping locations (see Table 8.1). Therefore, it appears to be rational for the North African decision makers to implement flexible cooperative allocation policies. These findings potentially provide valuable input to the recently established tri-national consultation mechanism with regard to the issues related to transboundary resource management. For each country, the calculated monetary gains from optimal management provide a strong incentive for cooperation (Siegfried and Kinzelbach, 2006).

Gains also result from intelligent scheduling of pumping, i.e. intermittent pumping between different boreholes or different aquifers. Over the optimisation time horizon of 50 years, demand can be covered without a large-scale renewal of groundwater provision infrastructure. Yet, boreholes tapping the deep IC aquifer will have to be relocated within the central basin due to the lowering of the piezometric head beyond the economic scarcity depth.

Because of the threat of salinisation in the Chott region of the central NWSAS basin, transboundary cooperation between Algeria and Tunisia is beneficial. However, large volume transboundary water transfers do not result from optimisation because of the large spatial extent of the resource. Each country has enough options to cover its demand in a cost-minimising manner on its own territory during the next 50 years. This finding does not discard the necessity for transboundary management. On the contrary, any policy deemed optimal for one country is only optimal as long as the others follow their corresponding optimal policies in the realisation because of the propagation of the pressure field. This becomes even more significant over the long time horizon. Finally, it has to be said that in the long run no sustainable solution exists at present pumping rates. Prices for water will increase indefinitely and a new strategy of water allocation is required. This can be achieved either by water desalination or a strong reduction of irrigation water demand, possibly by reducing crop production. The optimisation of water prices within a 50-year period is a means to provide time for the switch to a new type of economic activity.

8.4 UPSTREAM–DOWNSTREAM RELATIONS AND THE THREAT TO ECOSYSTEMS

8.4.1 General situation

Consumptive water use in the upstream of rivers leads to a lack of water in the downstream. The dramatic decrease of flows of even large rivers shows in the Amu Darya and Syr Darya basins, where upstream irrigation of cotton led to the partial drying up of the Aral Sea. The Yellow river, which traditionally was perennial, became an ephemeral river in the 1990s. The storage capacity of dams along the river is larger than the average annual flow volume. In 1997, 226 no-flow days over a stretch of 700 km were recorded (Xu, 2004). Similar fates await other large rivers such as the Nile.

The decrease of flows is not only of relevance to the downstream consumer sectors but also for the ecosystems. Wetlands which require yearly seasonal flooding are already destroyed or threatened by decreasing flows and more uniform flow regimes. The coastal wetlands in China have been reduced in area by 70%. But also riverine forests suffer from decrease of flows. An example is the case of the Populus Euphratica forests along the Tarim river in Xinjiang, China (see Figure 8.3 below). While these hardy trees can withstand years of drought, germination of seeds needs flooding. The upstream consumptive use on the Tarim has led to a substantial decline of the downstream forests.

All wetlands are important habitats for plants, animals and micro-organisms. Their importance to biodiversity and to humans is undisputed. Wetlands buffer hydrological extremes such as floods and droughts. Excess water during periods of rainfall is absorbed and the risk of floods is reduced. In periods of drought, wetlands gradually release water and ensure a minimal amount of available water in periods of little or no rainfall. The global area of wetlands diminished in the last century. Schutter (2003), for example, estimated that half of the world's

wetlands have disappeared since 1900, with reductions mainly in Europe and Northern America in the first half of the twentieth century and in Asia and Africa in the second half. There is a growing awareness that the remaining wetlands have to be protected. One of the most important conventions in this context is the 'Convention on Wetlands', signed in Ramsar, Iran, in 1971. It is an intergovernmental treaty which provides the framework for national action and international cooperation for the conservation and wise use of wetlands and their resources.

8.4.2 Example: The Okavango delta

The Okavango wetlands, commonly the called the Okavango delta, are located in northern Botswana. They are a large terminal wetland system spread out in the semi-arid Kalahari. Waters forming the Okavango river originate in the highlands of Angola, flow southwards until crossing the Namibian Caprivi Strip and eventually spread into the wetlands on Botswanan territory (Figure 8.2). Whereas the climate in the headwater region is subtropical humid with an annual precipitation water of up to 1300 mm, it is semi-arid in Botswana. High potential evapotranspiration rates cause practically all of the incoming water to be lost to the atmosphere together with the local precipitation water in the delta of 450 mm per year.

The Okavango wetlands are the only perennial water body in the Kalahari. They are crossed by migration routes of wildebeest, zebra and other animals. Besides the presence of water as such, it is the variability of the hydrological situation over the year which makes these wetlands unique. The combination of a highly seasonal inflow and local dry and wet seasons result in an ever-changing flooding pattern. A multitude of different environments and ecological niches have developed accordingly. The biodiversity found in the wetlands is large and can be primarily attributed to the special hydrological setting. Ramberg *et al.* (2006) list the number of identified species as 1300 for plants, 71 for fish, 33 for amphibians, 64 for reptiles, 444 for birds and 122 for mammals. With upstream activities still minor, the Okavango presently remains one of the largest, virtually pristine river systems on the African continent. It is included in the Ramsar list of wetlands of international importance (UNESCO, 1971). A number of future threats to the wetlands are apparent. They are linked mainly to the development of the upstream basin and eventually climate change.

The Okavango river basin is an international basin, with conflicting interests of the riparian states. Angola has large potentials both for hydropower and irrigated agriculture. The land area suitable for irrigation is estimated at 104 000 ha (Diniz and Aguiar, 1973, in Anderson *et al.*, 2006). Agricultural intensification and its detrimental impact on the Okavango delta with respect to both water quantity and quality are to be expected. Namibia suffers from severe water scarcity and is highly interested in the Okavango waters.

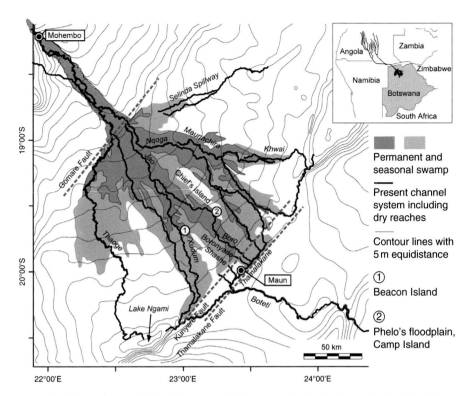

Figure 8.2. The Okavango wetland (commonly called the Okavango delta) covering the Panhandle and parts of the Okavango fan. The extent of the permanent and seasonal swamps is plotted from data of the Sharing Water Project (RAISON, 2004). The wetland section to the northwest of Gomare Fault is referred to as the Panhandle.

For Botswana the wetlands generate a considerable income through tourism. In 1996 tourism contributed 4.5% to GDP (Mbaiwa, 2005) with a large part being derived from the delta.

Bauer *et al.* (2006) developed a physically based numerical model describing the annual flooding of the delta which was developed further by Milzow *et al.* (2009). This model can be used to assess the impact of changes in upstream water allocation and the resulting change in flow regime. This includes a reduction in flow as well as changes in the temporal distribution of inflows by dams. It also allows to predict the consequences of hydraulic engineering projects within the delta itself such as dredging or cutting of papyrus.

Based on this model, diverse scenarios of surface-water and groundwater abstractions were simulated. It was found that the influence of domestic abstractions is negligible even when considering water needs for the entire basin and population growth as projected for 2025. An estimated $5\,000\,000$ m^3 per day of possible abstractions for irrigated agriculture are by one order of magnitude larger than domestic water needs. Abstractions of this magnitude are found to have a significant impact on the flooded area, especially for peripheral areas of the delta. Also, the decrease in flooded area is over-proportional to the decrease in inflow. In the 20 years time series simulated, the main impact was on the frequency of occurrence of small flooded areas. The extremely small flooded areas of less than 3000 km^2 increased

eight-fold. For a hypothetical dam, which was proposed at the upstream site of Popa Falls with a reservoir volume of 10^7 m^3, virtually no influence on the flooding of the delta was found. But larger assumed dams in Angola led to a damping of the strongly seasonal inflow and to spatial changes of the flooded area. Upstream seasonal swamps were decreased in size whereas downstream seasonal swamps were increased. However, the impact is smaller than the one generated by projected agricultural abstractions. Dredging of a specific channel – the Nqoga – is found to impact not only downstream regions of that channel but by backwater effects also to influence the water distribution at the scale of the entire delta.

The model in its latest form was used in combination with results from five general circulation models to assess the impact of climate change (Burg, 2007, Milzow *et al*, 2010). The baseline period chosen was from 1960 to 1990. While four models predicted drier future conditions, one model suggested wetter conditions. Assuming that the climate will more probably develop to drier conditions, the inflow to the delta is diminished. Together with abstractions for irrigation this would result in the largest detrimental effect on the delta with possibly up to 70% reduction of the seasonally flooded area in the worst case.

In order to establish the connection between simulated hydrology and status of the ecosystems, the correlation between the model output of the unchanged baseline period and vegetation classifications was analysed. A surprisingly good correlation was found between the simulated depth to groundwater and the classification into 12 eco-regions by McCarthy *et al.* (2006), while the flooding frequency was not well correlated.

The key task for hydrological models of the system, which emerges in this context, is to help transform modifications in the flow regime into modifications of ecosystem goods and services. Ultimately, decision makers will be interested to know what effect the construction of an upstream reservoir or irrigation district is going to have on the wildlife in the Okavango delta, on fishing activities and on the tourism industry. Hydrological models will therefore have to be coupled with ecosystem models and economic models. An interdisciplinary effort is required to find out how ecosystem state variables depend on hydrological state variables. Decision makers and managers of other wetland systems face similar challenges and early research into the 'eco-hydrology' of wetlands has been reported in the literature (e.g. Loucks, 2006).

The Okavango Delta Management Plan project (ODMP) was the first systematic attempt to collect all relevant scientific knowledge and data and to set up procedures for the long-term management of the Okavango delta in order to 'integrate resource management for the Okavango Delta that will ensure its long-term conservation and that will provide benefits for the present and future well being of the people, through sustainable use of its natural resources' (Government of Botswana, 2007).

However, the most crucial issue for the sustainable management of the Okavango river system and the Okavango delta is the transboundary nature of the water resources system. There is no doubt that increased development activities in upstream Angola will have significant impacts on the flow regime and water quality in the Okavango river. Benefits derived from ecosystem goods and services in the Okavango delta should be equitably shared among the three countries. In the absence of such a benefit-sharing agreement, Namibia and Angola will have no incentive to contribute to the preservation of a next-to-natural flow and water quality regime in the Okavango river. The sharing of benefits is made difficult by the fact that the opportunity cost for the upstream seems larger than the direct use value of the downstream. Botswana has an estimated income of 176 million US$ per year from tourism to the delta. The value of a previously planned water abstraction for Windhoek amounts to about 37 million US$ per year in water fees. Agricultural water abstractions in Angola may lead to foreign exchange savings by maize production of as much as 49 million US$. Further, only 90 million US$ of the 176 million US$ revenue from tourism in Botswana go into the fiscal resources of Botswana, while half leaves the country as revenue to foreign investors. Possibly a solution requires funds from the outside world and/or embedding of the water issue into other tripartite issues, which would as a package lead to a satisfactory solution with regard to the

conservation of the delta. Funds from the outside for the service of saving the wetland would not be illogical as the conservation of biodiversity is a benefit for and service to the global community.

The challenge for sustainable water resources management in the Okavango river basin is still huge. The contribution hydrological models could make to the solution of the problem is to provide transparent trade-off information and scenario simulations.

8.5 SALINISATION OF SOILS BY INAPPROPRIATE IRRIGATION PRACTICE

8.5.1 General situation

Salinisation occurs naturally (primary salinisation) or due to human activities (secondary salinisation). A soil becomes saline if more salt is deposited than removed. Comprehensive overviews covering most of the causes, consequences and possibilities to tackle the problem of salinisation can be found in Hillel (2000), Jakeman *et al.* (1995) and Richards (1954), to mention a few. Besides deforestation, irrigation with river water is one of the most common causes of secondary salinisation. Application of surface water without adequate drainage causes a rise of the groundwater table. Salt stored in the subsoil is dissolved as the groundwater table rises. Once the depth to groundwater becomes small, capillary rise will lead to direct evaporation of groundwater. As the water evaporates, the dissolved salts accumulate at the soil surface. The connection between the depth to groundwater and the rate of soil salinisation requires coordination between aquifer management and agricultural management in order to arrive at a sustainable regime.

The rapidly increasing food demand during the last 50 years has forced nations worldwide to expand existing irrigation infrastructure. Soil salinisation has accompanied societies relying on irrigation throughout history (Hillel, 1990) and remains one of the major threats today. Soil salinisation reduces productivity of soils and in the extreme case leads to sterilisation.

There is an ongoing debate whether irrigated agriculture is sustainable or not. Hillel (2000) concluded that it can be sustainable, but at a cost. This cost includes investments in the drainage and irrigation infrastructure. Water-saving agriculture reduces the infiltration and consequently the rise of the groundwater table. A functioning drainage system removes excess irrigation water and allows keeping the groundwater table at a lower level. The conjunctive use of surface water and groundwater is another possibility of controlling the groundwater table. However, pumping groundwater from an underlying aquifer system can result in the intrusion of saline water and a decrease of irrigation water quality in the long term. Khan *et al.* (2006) concluded that

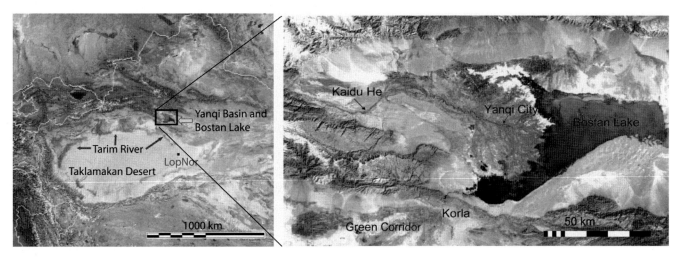

Figure 8.3 (Plate 8). Location of the Kaidu–Kongque system and the Tarim river basin.

even the modern, high-tech irrigation zone of the Murrumbidgee region (Australia) was not sustainable in terms of soil salinisation. Considering the rapidly growing demand for food on a global scale, it is unlikely that investments to control soil salinity on a large scale will be provided fast enough to effectively slow down the loss of agricultural land due to soil salinisation. It is estimated that 80 million hectares out of 240 million hectares of irrigated land are more or less affected by salinisation.

The experience described above shows that in arid regions agricultural water management is inseparable from salt management if sustainability is to be achieved. Water is the carrier of dissolved salts and, depending on where it evaporates, the salt is deposited in inappropriate places.

8.5.2 Example: The Yanqi basin in Xinjiang, China

A typical example for a salinisation problem is the Kaidu–Kongque river system in Xinjiang, China (Figure 8.3). The Kaidu river originates in the Tian Shan mountains and flows through the Yanqi basin into Bostan lake, leaving the basin under the name of Kongque river. The Kongque river supplies the so-called Green Corridor between Korla and the now dried-out Lop Nor with water. Before irrigation in the Yanqi basin and along the Kongque river took place, Lop Nor was the final sink of the Kaidu–Kongque system.

The rapid population growth in the Yanqi basin since the 1950s has led to a non-sustainable environmental situation. With an annual precipitation as low as 20 mm per year crop production is only feasible with full irrigation. Irrigation systems based on water drawn from the Kaidu were introduced, transforming the basin into one of the most productive agricultural areas in Xinjiang. As a consequence the groundwater table rose, mobilising salt deposited in the unsaturated zone. In many places the distance to groundwater fell below 1 to 2 metres, under which

conditions the capillary rise led to direct evaporation of groundwater and the deposition of all dissolved salts in the topsoil. The increasing salinity can be controlled by leaching the soil. However, in the long run, this practice only worsens the situation, as the infiltrating leaching water additionally raises the groundwater table, increasing the direct evaporation of groundwater and therefore the leaching requirement for the coming irrigation cycles. The direct evaporation of groundwater is driving soil salinisation in the Yanqi basin and reduces the available water resources for the downstream. Today, some 60% of the cultivated area in the basin is strongly affected by salinisation (Dong *et al.*, 2001).

At the same time, the flow of Kaidu water into the lake was reduced. The salt concentration of the remaining water resources increased due to the high evaporation losses upstream. Today the high salinity of the inflow is the most serious threat to the lake ecosystem.

Several options exist to manage the Yanqi basin in a more sustainable way. Reducing the irrigated area, improving the efficiency of the irrigation system or planting salt-tolerant crops would help to save water and to increase the depth to groundwater. One of the most promising options is, however, the substitution of part of the irrigation water drawn directly from the river by river water abstracted indirectly from the aquifer system. Pumping groundwater lowers the groundwater table, reduces unproductive phreatic evaporation and therefore slows down the salinisation of topsoil. Groundwater is hardly used today, as due to energy requirements it is more expensive than surface water. However, if the water table can be kept low by pumping groundwater, the conservation of soil for continued agricultural use along the Kaidu river and the benefits of a larger flow to the downstream ecosystems might strike a balance with a higher price of water. Also, all water-saving irrigation methods such as drip irrigation are much more easily applied using conveniently controlled and sediment-free groundwater than surface water. Finally drip irrigation allows considerable savings of fertilizer as it can be added to the irrigation water in dissolved form. The major decision variable to steer the system into a desirable state without reducing the irrigated area is the ratio of irrigation water taken directly from the river to water abstracted from the aquifer, which of course also originates from the river by seepage.

To assess the feasibility of this option, a spatially distributed hydrological model simulating groundwater and surface water as well as the coupling between them was constructed (Brunner, 2005). Examples for spatially distributed parameters used in the model are a digital elevation model obtained from SAR-images and the distribution of infiltration rates under the irrigated fields. An important data set required in this task was the spatial distribution of evapotranspiration, which was calculated from NOAA-AVHRR images according to a method proposed by Roerink *et al.* (2000). The verification data for the model consisted of observations of the groundwater table, the spatial distribution of phreatic evaporation and the distribution of salt in the project area Brunner *et al.* (2008), Brunner *et al.* (2007), Li *et al.* (2009) as well as observed infiltration rates along the rivers. Both the phreatic evaporation and the distribution of salt in the top soil layer were obtained from satellite remote sensing data. Ground truth was obtained with geophysical methods (TEM), DGPS and through stable isotopes, indicating evaporation. The model reproduced the verification data satisfactorily and predicts that the substitution of every 1 m^3 of river water by pumped groundwater will increase the available downstream resources by at least 0.75 m^3. The scenarios show that without reducing the agricultural production, sustainable solutions for the water allocation in the Yanqi basin and its downstream areas exist.

The natural system around Bostan lake is accumulating salts. This happens in the salt marshes along the lake shore (see Figure 8.3), which are the long-term sink of practically all salts mobilised in the basin. A sustainable management of the basin has to maintain this natural sink of salt. An excessive groundwater abstraction would impair the functioning of this sink by inverting the gradient between the marsh and the aquifer. Therefore the amount of groundwater pumped for irrigation has to be chosen judiciously.

8.6 POLLUTION OF WATER BODIES OF LARGE RESIDENCE TIME WITH PERSISTENT OR RECYCLABLE POLLUTANTS

Worldwide river and lake pollution pose immense problems and the present state is definitely not sustainable. Still we refrain from including river pollution into the more fundamental problems of sustainability. The reason for that is that in principle polluted rivers can be rehabilitated within times on the order of a decade. The residence time of water in the river Rhine, for example, between Lake Constance and the mouth is just 8 days. An instantaneous closing down of pollution sources could therefore restore water quality within that short time span. The successful rehabilitation programmes of the Rhine and the Thames have shown that river pollution is not irreversible. The relatively small pre-alpine lakes in Switzerland, which were plagued by eutrophication in the 1970s, were brought back to an oligotrophic state within 15 years after the installation of wastewater treatment plants with phosphorous elimination.

The situation is different in water bodies with a large residence time. Typically, larger aquifers have residence times of hundreds to thousands of years. A pollution with persistent contaminants will remain in the system for an accordingly long period. Remediation is extremely expensive and often extremely inefficient.

Large lakes also have large residence times, eventually prolonged by density stratification. In addition, if deposits of phosphorous are already large as in some lakes in China (Dianchi or Taihu), the time for rehabilitation is still huge as algal blooms are triggered to a large part by recirculating phosphorous which is still present even if external sources are switched off. Similarly, inland seas like the Baltic or the Mediterranean, which have been sinks for recyclable or persistent pollutants for many years, are difficult to clean up. As clean-up in all described water bodies of long residence time takes more than one generation, their pollution is quasi-irreversible and therefore touches the issue of sustainability.

The most common persistent contaminants of groundwater bodies are nitrate (under aerobic aquifer conditions), persistent organic pollutants such as chlorinated hydrocarbons and MTBE, some pesticides and salts. One might expect that among the organic pollutants chlorinated hydrocarbons are the most important due to their high persistence. However, chlorinated hydrocarbons are mainly a problem in industrialised countries while globally the most prominent pollutant is salt, especially in arid regions and coastal areas, where seawater intrusion occurs. Seawater intrusion is caused by pumping of groundwater from a coastal aquifer. Excessive pumping reduces piezometric heads and allows the naturally forming salt wedge to extend further inland. Well-known examples are Israel's Coastal Plain aquifer (Nativ, 2004) or the coastal aquifers in India, China, California and Spain. These systems have been pumped to cover irrigation water demand. As a consequence of the ongoing excessive use of groundwater, seawater has propagated inland, reaching the irrigation wells. Rising seawater levels will also increase seawater intrusion.

Another problem perceived as a future threat is the increasing abstraction of deep groundwater systems. Due to their age, these are not polluted but continued pumping will bring shallow, polluted groundwater down to these wells. An increasing salinity of pumped, deep groundwater has been reported, for example, for the Shepparton region in Australia (Prendergast *et al.*, 1994), where deep groundwater was used conjunctively with surface water to cover the irrigation water demand.

8.7 TOOLS FOR DECISION SUPPORT

The analysis of sustainability is based on mechanistic relations between stresses and outcomes. When carried out in a quantitative way, this analysis consists of building a model of the region, its hydrological behaviour and the coupling of hydrological variables to other relevant system parameters. All models require data and that turns out to be their weak point: if no data are available, models are meaningless. It is not sufficient to have a rather complete set of input data. It is also necessary to have some monitoring data for the past, which can serve as a basis for calibration and/or verification of the model.

Water resources assessment, modelling and management are hampered considerably by a lack of data, especially in semi-arid and arid environments with a weak observation infrastructure. Usually, only a limited number of point measurements are available, while hydrological models need spatial and temporal distributions of input and calibration data. If such data are not available, models cannot play their proper role in decision support as they are notoriously underdetermined and uncertain. Recent developments in remote sensing have opened new sources for distributed spatial data. As the relevant entities such as water fluxes, heads or transmissivities cannot be observed directly by remote sensing, ways have to be found to link the observable quantities to input data required by the model. The main possibilities are: (a) use of remote sensing data to create some of the spatially distributed input parameter sets for a model, and (b) constraining of models during calibration by spatially distributed data derived from remote sensing. In both ways models can be improved conceptually and quantitatively (Brunner et al., 2007b). One example is used for illustration.

For flat terrain the groundwater recharge potential over long time intervals is the long-term average residual between precipitation P and evapotranspiration ET. Both quantities can be estimated from remote sensing data. The precipitation can be estimated from cloud temperature data by certain algorithms (e.g. Herman et al., 1997) in combination with precipitation data from meteorological stations on the ground. The Famine Early Warning Systems Network (http://www.cpc.noaa.gov/products/fews/) offers such data at a daily temporal resolution for all of Africa, South and Southeast Asia, Central America and the Caribbean, and Afghanistan. Evapotranspiration can be derived from multispectral satellite data via a surface energy balance. To put it simply, a dry pixel will heat up to higher temperatures than a pixel which has a large amount of water available for evaporative cooling. In this sense radiation data can be related to evapotranspiration. The fraction of net radiation energy consumed by evaporating water can be estimated with different methods. In SEBAL (Surface Energy Balance Algorithm for Land; Bastiaanssen et al., 1998a, 1998b), the energy fluxes in the surface energy balance are calculated explicitly, while in a simplified method described by Roerink et al. (2000) this fraction is determined from a pixel-wise plot of surface temperature versus albedo. Other methods use different dimensions of the feature space instead, e.g. the Normalised Difference Vegetation Index (NDVI), which is a measure of the vigor of vegetation growth (Sandholt and Andersen, 1993). Unfortunately, both ET and P obtained from remote sensing are inaccurate. Calculating the difference, P – ET, leads to error propagation, especially when both quantities are of similar magnitude. This is often the case in semi-arid and arid areas. Still, the spatial patterns of P–ET may be of help in regionalisation of traditional point measurements of recharge. The pattern information can be calibrated with point measurements, for example obtained with the chloride method (Brunner et al., 2004).

Remote sensing, be it from an airplane or a satellite platform, opens up new avenues for water resources management especially in large countries with weak infrastructure. Local hydrological research can be interpolated and completed into more areally exhaustive data sets. Generally, remote sensing data are always of interest if they reduce the degrees of freedom of a model by an objective zonation of the whole region considered. We use satellite images not only in the determination of precipitation and evapotranspiration. From the vegetation distribution and the distribution of open water surfaces after the rainy season hints to the spatial distribution of recharge can be abstracted. Multi-spectral satellite images can show the distribution of salts at the soil surface. For large areas the GRACE mission allows soil and ground water balances to be calculated from variations in the gravitational field of the Earth. New developments in geomatics allow us to obtain accurate digital terrain models, on small scales from laser scanning or stereophotometric methods and on a large scale from radar satellite images. In both cases checkpoints on the surface are required which can be obtained with differential GPS. Digital terrain models are important in all cases where the distance to groundwater table or a flooded area at varying water surface elevation play a role. The combination of remote sensing, automatic sensors with data recorders, GPS and environmental tracers in combination with the classical methods of hydrology and hydrogeology can significantly improve water resource studies, especially those cases where, in earlier times, lack of infrastructure and poor accessibility were limiting.

Despite the widespread use of models, care is advisable. Models are notoriously uncertain and their prognostic ability is often overestimated. The necessary input data are usually only known partially or at certain points, instead of in their spatial distribution. Lacking input, parameters have to be identified during calibration, which as a rule does not yield a unique solution. Hydrological models are notoriously overparametrised and good model fits are feasible with different sets of parameters, which in a prognostic application of the model, however, would yield different results. A responsible use of models acknowledges their uncertainty and takes it into account when interpreting results. As illustrated in Chapter 4, model uncertainty estimation methods available today go far beyond the classical engineering philosophy of best case–worst case analysis.

A model can be used to test whether a water allocation strategy is sustainable. One way of doing so is to let the model run towards steady state (assuming that all boundary conditions such as external drivers are stationary). If the steady state exists and is acceptable from ecological and socio-economic viewpoints, the strategy is sustainable. In reality the boundary conditions may not be constant and systematic changes (e.g. climate change) may be involved. In that case an adaptive strategy is required which follows the changing trend.

The usefulness of models need not suffer from the admission that they are uncertain. The resource engineer is not interested in every detail of nature. He or she wants to propose a robust solution, which will function even if the system does not end up behaving exactly as was anticipated. For this type of decision the stochastic approach is appropriate as it allows quantification of the risk of making a wrong decision.

Decisions on sustainability take place in the economic–political space. Therefore the usefulness of groundwater models is strongly dependent on their ability to interface with socio-economic aspects, such as cost and benefits.

The aquifer exploitation problem is a typical common pool problem. Consider the situation of n users of an aquifer, which receives a total recharge Q_N. As long as the pumping rate of the n users is smaller than Q_N, the difference between recharge and consumptive use will flow out of the basin to a receiving stream. It is clear that in the long term the total abstraction cannot surpass the recharge rate. Sooner or later the abstraction has to adapt to the availability of water. It is, however, wise to get to that equilibrium at an early stage, which means at a relatively high water table. Higher water tables allow larger amounts of short-term overpumping during droughts. Moreover, higher water tables reduce the specific pumping cost for all users. If every user egoistically opts for a large drawdown and a deep well, each will experience the damage of higher specific pumping costs. Groundwater is a resource which rewards partnership and punishes egoism. Negri (1989) showed how a collective decision in the common pool problem leads to a better solution than the sum of all independent, individual decisions.

In the soil salinisation problem, the situation is similar only with the opposite sign in the water table change. All irrigators infiltrate water and cause a groundwater table rise, which leads to evaporation from the groundwater table and thus the initiation of salinisation. Collective self-restraint would again lead to a better solution.

8.8   THE CONNECTION TO DECISION MAKERS

There are many ways in which scientific knowledge finds its way into practical decision making. Unfortunately, the usual way of dissemination of scientific knowledge by publication in SCI-listed international journals is not very efficient. In our own projects three different approaches were chosen. They all involved close cooperation with national or regional water administration offices. In the North West Sahara aquifer system example, we took part in a common project of the three countries involved. Officers from all three countries were instructed in groundwater modelling and recharge estimation. A fruitful cooperation with a scientist at a university of the region was established (Zammouri *et al.*, 2007). The results of the research were discussed with the highest water officials of the three countries. Via this pathway further useful discussions between the competing countries were initiated. No direct and concrete result is visible yet. However, the principles of exploitation of a non-renewable resource will leave their mark on future debate.

In the case of the Okavango delta, an officer of the Department of Water Affairs participated in the research as a PhD student, through whom the influx of data from Botswana as well the backflow of results to Botswana was guaranteed. The results are expected to have a strong impact on future negotiations between the riparian states, concerning the allocation of the Okavango water resources organised within the framework of the Permanent Okavango River Basin Commission (OKACOM).

For the Yanqi basin project, the numerical model was translated into a simulation game, SIMSALIN. In this game the player is either a water manager or an agricultural planner. The following decisions have to be made for four irrigation districts:

- choice of cultivated area;
- choice of crop mix;
- choice of irrigation method and pre-irrigation requirements;
- choice of ratio of contributions from the alternative water resources (surface water or groundwater, groundwater from first or second aquifer);
- choice of maintenance for drainage canals;
- choice of lining of irrigation canals;
- choice of water price.

The model calculates for 10-year periods the outcome of the decisions with respect to four criteria:

- net benefit of agricultural production (the higher the better);
- area with depth to groundwater smaller than 2 m (danger of salinisation the smaller the better);
- salinity of the lake water (the lower the better);
- outflow to the 'Green Corridor' of Populus Euphratica forests in the downstream (the higher the better).

These are then used to judge the quality of the management choices made.

A composite of the four quantities is used to calculate a score. If the allocated water surpasses the amount of river water present, the model stops simulating. After every 10-year interval, changes in allocation can be made and the player can learn from his/her past experience. The game will be used in the education of Chinese water administrators. We hope that in this way more awareness of cause–effect relations in a complex system can be created and regions will be analysed in a more holistic fashion.

8.9   CONCLUSIONS

All of the water-related problems discussed above are a direct consequence of the ever-increasing global population. According to an estimate of the United Nations (UN), the world population was around 6.5 billion in 2005. In 2050, it could be close to 9 billion. The pressure on available resources will increase and there is no doubt that the need for sustainable practices will increase. Finding the balance between immediate needs to supply the world with essential goods such as food at the lowest possible cost and the long-term need for future generations will become more and more difficult.

Modern tools of hydrological science such as modelling, remote sensing and geophysics are helpful especially in regions with weak infrastructure to quantify the implications of human interaction and to give advice to decision makers on the sustainability of water management practices (Brunner et al., 2007). Such models summarise the status quo, increase system understanding and represent the only means to make predictions. They are bound to be crude and simulations will always be idealised. Still, they can serve as points of reference. Stochastic methods allow the incorporation of at least some degree of uncertainty which is necessary to find conservative or robust solutions (Franssen et al., 2004, 2008). A further common feature is that sustainable solutions require the system boundaries to be sufficiently large, often transgressing political boundaries. While science can give some decision support, the decisions for or against sustainability are made in the political arena. However, the growing global awareness of the need to protect water resources, the increasing understanding of hydrological systems and experience with different management practices allow us to retain some optimism.

REFERENCES

Bastiaanssen, W. G. M., Menenti, M., Feddes, R. A. and Holtslag, A. A. M. (1998a) A remote sensing surface energy balance algorithm for land (SEBAL). 1. *Formulation. Journal of Hydrology* 213(1–4), 198–212.

Bastiaanssen, W. G. M., Pelgrum, H., Wang, J. et al. (1998b) A remote sensing surface energy balance algorithm for land (SEBAL). 2. Validation. *Journal of Hydrology*, 213(1–4), 213–229.

Bauer, P., Gumbricht, T. and Kinzelbach, W. (2006) A regional coupled surface water/groundwater model of the Okavango Delta, Botswana. *Water Resour. Res.* 42, W04403.

Binmore, K. (1994) *Game Theory and the Social Contract.* Vol. 1. MIT Press.

Brunner, P. (2005) *Water and Salt Management in the Yanqi Basin*, China. IHW. PhD Thesis. ETH Zurich No.16210, ISBN 3–906445–26–7, http://e-collection.ethbib.ethz.ch/show?type=diss&nr=16210.

Brunner, P., Bauer, P., Eugster, M. and Kinzelbach, W. (2004) Using remote sensing to regionalize local precipitation recharge rates obtained from the chloride method. *Journal of Hydrology* 294(4), 241–250.

Brunner, P., Li, H. T., Kinzelbach, W., Li, W. P. (2007a) Generating soil electrical conductivity maps at regional level by integrating measurements on the ground and remote sensing data. *Int J Remote Sens*, 28(15), 3341–3361.

Brunner, P., Franssen, H. J. H., Kgotlhang, L., Bauer-Gottwein, P. and Kinzelbach, W. (2007b) How can remote sensing contribute in groundwater modeling? *Hydrogeology Journal* 15, 5–18.

Brunner, P., Li, H. T., Kinzelbach, W., Li, W. P. and Dong, X. G. (2008) Splitting up remotely sensed maps of evapotranspiration into maps of evaporation and transpiration. *Water Resour Res*, 44 W08428, doi:10.1029/2007WR006063.

Burg, V. (2007) *Climate Change Affecting the Okavango Delta.* Diploma Thesis, July 2007; ETH, Eidgenössische Technische Hochschule Zürich, Institute of Environmental Engineering (IfU). http://e-collection.ethbib. ethz.ch/show?type=dipl&nr=338. Cited 27 November 2007.

Diniz, A. C. and Aguiar, F. P. (1973) Recursos em terras com aptidão para o regadio da bacia do Cubango, Série Técnica, no. 33, Instituto de Investigação Agronómica de Angola, Nova Lisboa, Angola, 1973 (in Portuguese).

Dong, X., Jiang, T. and Jiang, H. (2001) Study on the pattern of water resources utililzation and environmental conservation of Yanqi Basin. In *Development, Planning and Management of Surface and Groundwater Resources*, ed. G. Li, 333–340. IAHR Congress proceedings. Tsinghua University Press, Beijing, China.

FAO (2006) *Food and Agriculture Statistics Global Outlook.* http://faostat. fao.org/Portals/_Faostat/documents/pdf/world.pdf. Statistics Division of the FAO.

Franssen, H. J. H., Stauffer, F. and Kinzelbach, W. (2004) Joint estimation of transmissivities and recharges – application: stochastic characterization of well capture zones. *Journal of Hydrology* 294, 87–102.

Franssen, H. J. H., Brunner, P., Makobo, P. and Kinzelbach, W. (2008) Equally likely inverse solutions of a groundwater flow problem including pattern information from remote sensing images. *Water Resour. Res.* doi: 10.1029/2007WR006097.

Gleick, P. H. (1993) *Water in Crisis: A Guide to the World's Fresh Water Resources.* Oxford University Press.

Government of Botswana (2007) *Okavango Delta Management Plan Project: Draft Final Okavango Delta Management Plan.* Department of Environmental Affairs, Gaborone, Botswana.

Herman, A., Kumar, V. B., Arkin P. A. and Kousky, J. V. (1997) Objectively determined 10-day African rainfall estimates created for famine early warning systems. *Int. J. Rem. Sens.* 18, 2147–2159.

Hillel, D. (1990) *Out of the Earth: Civilization and the Life of the Soil.* Free Press.

Hillel, D. (2000) Salinity management for sustainable irrigation: integrating science. *Environment and Economics.* World Bank.

Jakeman, A. J., Nix, H. A. and Ghassemi, F. (1995) *Salinisation of Land and Water Resources: Human Causes, Extent and Management.* CAB International.

Khan, S., Tariq, R., Cui, Y. L. and Blackwell, J. (2006) Can irrigation be sustainable? *Agricultural Water Management* 80, 87–99.

Li, H. T., Brunner, P., Kinzelbach, W., Li, W. P. and Dong, X. G. (2009) Calibration of a groundwater model using pattern information from remote sensing data. *Journal of Hydrology*, 377(1–2), 120–130.

Loucks, D. P. (2006) Modeling and managing the interactions between hydrology, ecology and economics. *J. Hydrol.* 328, 408–416.

Mbaiwa, J. E. (2005) Enclave tourism and its socio-economic impacts in the Okavango Delta, Botswana. *Tour. Manag.* 26, 157–172.

McCarthy, T. S. (2006) Groundwater in the wetlands of the Okavango delta and its contribution to the structure and function of the ecosystem. *J. Hydrol.* 320, 264–282.

Milzow, C., Kgotlhang, L., Kinzelbach, W., Meier, P. and Bauer-Gottwein, P. (2009) The role of remote sensing in hydrological modelling of the Okavango Delta, Botswana. *J. Environ. Manag.* 90(7), 2252–60.

Milzow, C., Burg, V., Kinzelbach, W. (2010) Estimating future ecoregion distributions within the Okavango Delta Wetlands based on hydrological simulations and future climate and development scenarios. *Journal of Hydrology* **381**(1–2), 89–100.

Nativ, R. (2004) Can the desert bloom? Lessons learned from the Israeli case. *Ground Water* **42**, 651–657.

Negri, D. H. (1989) The common property aquifer as a differential game. *Water Resour. Res.* **25**, 9–15.

OSS (Observatoire du Sahara et du Sahel) (2003) *The North West Sahara Aquifer System: Joint Management of a Transborder Basin.* Observatoire du Sahara et du Sahel.

Postel, S. (1997) *Last Oasis.* W. W. Norton.

Prendergast, J. B., Rose, C. W. and Hogarth, W. L. (1994) Sustainability of conjunctive water-use for salinity control in irrigation areas – theory and application to the Shepparton Region, Australia. *Irrigation Science* **14**, 177–187.

Ramberg, L., Hancock, P., Lindholm, M. *et al.* (2006) Species diversity of the Okavango Delta, Botswana. *Aquat. Sci.* **68**, 310–337.

Richards, L. A. (1954) *Diagnosis and Improvement of Saline and Alkali Soils.* US-SalinityLab.

Roerink, G. J., Su, Z. and Menenti, M. (2000) S-SEBI: A simple remote sensing algorithm to estimate the surface energy balance. *Phys. Chem. Earth (B)* **25**, 147–157.

Sahagian, D. L., Schwartz, F. W. and Jacobs, D. K. (1994) Direct anthropogenic contributions to sea-level rise in the 20th century. *Nature* **367**(6458), 54–57.

Sandholt, I. and Andersen, H. S. (1993). Derivation of actual evapotranspiration in the Senegalese Sahel, using NOAA-AVHRR data during the 1987 growing-season. *Remote Sensing of Environment* **46**(2), 164–172.

Schutter, J. D. (2003) Wetlands management. In *Water Resources and Environment*, ed. R. Livernash. Technical Note G.3. Washington http://iucn.org/places/medoffice/cdflow/conten/5/pdf/5–3-International-Guid/World-Bank-ENG/Waterbody-Management/NoteG3Waterbody.pdf, The World Bank.

Siegfried, T. (2004) *Optimal Utilization of a Non-Renewable Transboundary Groundwater Resource – Methodology, Case Study and Policy Implications.* PhD Thesis, http://e-collection.ethbib.ethz.ch/show?type= diss&nr= 15635, Zurich, Switzerland.

Siegfried, T. and Kinzelbach, W. (2006) A multiobjective discrete stochastic optimization approach to shared aquifer management: methodology and application. *Water Resour. Res.* **42**(2).

UN (2006) *World Population Prospects: The 2006 Revision and World Urbanization Prospects: The 2005 Revision*, Population Division of the Department of Economic and Social Affairs of the United Nations Secretariat, http://esa.un.org/unpp.

UNESCO (1971) *Convention on Wetlands of International Importance especially as Waterfowl Habitat.* Ramsar (Iran), 2 February 1971. UN Treaty Series No. 14583. As amended by the Paris Protocol, 3 December 1982, and Regina Amendments, 28 May 1987.

Xu, J. X. (2004) A study of anthropogenic seasonal rivers in China. *Catena* **55**, 17–32.

Zammouri, M., Siegfried, T., El-Fahem, T., Kriâa, S. and Kinzelbach, W. (2007) Salinization of groundwater in the Nefzawa oases region, Tunisia: Results of a regional-scale hydrogeologic approach. *Hydrogeology Journal* doi 10.1007/s10040–007–0185-x.

Index

Printed in the United States
By Bookmasters